PROBLEM-SOLVING EXPERIENCES IN

RANDALL I. CHARLES
ROBERT P. MASON
LINDA MARTIN

GRADE
4

MATHEMATICS

TEACHER SOURCEBOOK

ADDISON-WESLEY PUBLISHING COMPANY

Menlo Park, California • Reading, Massachusetts • Don Mills, Ontario
Wokingham, England • Amsterdam • Sydney • Singapore
Tokyo • Madrid • Bogotá • Santiago • San Juan

This book is published by the Addison-Wesley Innovative Division.

Design: Betsy Bruneau Jones

Illustrations: Shirley Bortoli
Elizabeth Callen
Jane McCreary

The blackline masters that accompany this publication are designed to be used with appropriate duplicating equipment to reproduce copies for classroom use. Addison-Wesley Publishing Company grants permission to classroom teachers to reproduce these masters.

Published simultaneously in Canada.

ISBN-0-201-20831-8 (hard cover)
ISBN-0-201-20294-8 (soft cover)
11 12 13 14 15 - ML - 95 94

Contents

Overview

PROBLEM-SOLVING EXPERIENCES IN MATHEMATICS, GRADE 4, is an instructional program designed to supplement any basal textbook. The program consists of 150 problem-solving experiences and a teaching strategy for problem solving. A separate package of blackline masters provides statements of the problem-solving experiences to be duplicated for students. This book and its blackline masters are part of a series of eight, one for each of grades 1 through 8.

All of the problem-solving experiences in this book have been used with hundreds of children at this grade level to assure that they are interesting to students and that the difficulty of the experiences is appropriate for a wide range of student abilities. Furthermore, teachers' evaluations of problem difficulty were used to help select and sequence the problem-solving experiences.

Goals of the program

The ultimate goal of any problem-solving program is to improve students' performance at solving problems correctly. Although this is the ultimate goal, one's instructional goals need to be more specific and developmental. The specific goals of PROBLEM-SOLVING EXPERIENCES IN MATHEMATICS are to:

1. Improve students' willingness to try problems and improve their perseverance when solving problems.
2. Improve students' self-concepts with respect to their abilities to solve problems.
3. Make students aware of problem-solving strategies.
4. Make students aware of the value of approaching problems in a systematic manner.
5. Make students aware that many problems can be solved in more than one way.
6. Improve students' abilities to select appropriate solution strategies.
7. Improve students' abilities to implement solution strategies accurately.
8. Improve students' abilities to monitor and evaluate their thinking while solving problems.
9. Improve students' abilities to get more correct answers to problems.

Organization of the program

There are four types of problem-solving experiences provided by this program: *problem-solving skill activities, one-step problems, multiple-step problems,* and *process problems.* There are 150 problem-solving experiences grouped in 30 sets of five. Each set includes one skill activity, one one-step problem, one multiple-step problem, and two process problems.

Problem-solving skill activities are experiences intended to promote the development of thinking processes involved in problem solving. Ten types of skill activities are included in this program.

1. Given a story, write a question that could be answered using the data in the story.
2. Given a problem with unneeded data, identify the data needed to find a solution.
3. Given a problem with missing data, make up appropriate data for solving the problem.
4. Given a problem, identify the operation(s) needed to find a solution.
5. Given a problem without numbers, identify the operation(s) needed to find a solution.
6. Given a table or organized list, write a story problem that would be solved by making and using that table or list.
7. Given a picture, write a story problem that would be solved using the picture.
8. Given a number sentence, write a story problem that would be solved using that number sentence.
9. Given the numerical part of the answer to a problem, write the answer in a complete sentence.
10. Given a problem and its answer (written in a complete sentence), determine the reasonableness of the answer.

One-step problems and multiple-step problems require the student to translate story situations, mentally or on paper, into one or more number sentences. **One-step problems** are the familiar "story problems" that are a traditional part of school mathematics programs. The primary technique or strategy used to solve one-step problems is called *choose the operation.* That is, students must decide whether to add, subtract, multiply, or divide. The primary strategy used to solve **multiple-step problems** is *choose the operations.* Here the students must choose two or more of the operations (addition, subtraction, multiplication, or division) to find a solution.

At grades 1–8, **process problems** are ones that cannot be solved by simply adding, subtracting, multiplying, or dividing. Instead, process problems are solved using one or more of these strategies.

1. Guess and check.
2. Draw a picture.
3. Make an organized list.
4. Make a table.
5. Work backwards.
6. Look for a pattern.
7. Use logical reasoning.

Because process problems cannot be solved by simply choosing one or more operations, they exemplify the need for and provide practice with the thinking processes inherent in problem solving.

The following chart indicates what strategies can be used to solve the process problems in this book. The chart does not attempt to indicate all possible ways of solving the problems—only those that are most commonly used.

PROBLEM-SOLVING EXPERIENCES IN MATHEMATICS was designed with a "problem-of-the-day" approach in mind. The program provides a problem-solving experience for almost every day of the school year (150 experiences), and the problems were selected and sequenced so the concepts and skills needed to solve each problem would have been introduced to students approximately two months before they are encountered here, *if* the teacher follows the scope and sequence of lessons in most basal textbooks. For problems at the beginning of the year, concepts and skills are limited to those most students should have mastered at the previous grade level. This organization means that students' work is limited to a *review* of concepts and skills. The emphasis of problem-solving instruction can thus be on understanding problems, selecting and implementing appropriate solution strategies, and checking one's work, rather than on carrying out computational skills.

The one-step and multiple-step problems in this program were also selected so as to balance the computational skills needed to find solutions. That is, the solutions involve the use of addition, subtraction, multiplication, and division skills to a degree appropriate for this grade level.

Since the process problems in this book have been used with hundreds of children, we were able to determine (by analyzing the children's work) the strategies most students use to solve a particular problem. Using this information, we organized the 60 process problems in the following way (see page

Organization of Process Problems

Guess and Check	*Draw a Picture*	*Make a List*	*Make a Table*	*Work Backwards*	*Look for a Pattern*	*Use Logical Reasoning*
Problem	9, 10	14, 15	19, 20	24, 25	29, 30	34, 35
4, 5	44, 45	49, 50	54, 55	59, 60	64, 65	69, 70
39, 40	79, 80	75	64	119, 120	114	74
74, 75	85	80	90	130	125	144
89	94	84	104, 105		139	149
99	109, 110	95	114, 115		145	
129	135	99, 100	124, 125		150	
134	144	134, 135	129			
144	149		139, 140			
149			144, 145			
150			149, 150			

Set 1
Guess and check.

Set 2
Draw a picture.

Set 3
Make an organized list.

Set 4
Make a table.

Set 5
Work backwards.

Set 6
Look for a pattern.

Set 7
Use logical reasoning.

- The two process problems in each of sets 1–7 are ones most students will solve using the strategy shown for that set.
- Each process problem in sets 1–7 is accompanied by a hint. For example, in set 4, each problem is followed with, "Complete this table to help you solve this problem" (table is provided).

Set 8
Guess and check.

Set 9
Draw a picture.

Set 10
Make an organized list.

Set 11
Make a table.

Set 12
Work backwards.

Set 13
Look for a pattern.

Set 14
Use logical reasoning.

- The process problems in each of sets 8–14 are still grouped by the most probable solution strategy.
- The first process problem in each of sets 8–14 is accompanied by a hint, the second is not. The hints here are more general than those in the first seven sets.

Set 15

Set 16

Set 17

.

.

.

Set 30

- The two process problems in each of sets 15–30 are usually mixed, that is, students would most likely not use the same solution strategy on each.
- Hints no longer accompany problem statements.

The two process problems in each of the first 14 sets are matched by a probable solution strategy and accompanied by hints to enable you to teach students how to use problem-solving strategies *if your students need this type of direct instruction.* If students are new to problem solving, it is quite likely that they will need some direction concerning, for example, how to make and complete tables. Even though the process problems are matched within a set for the first 14 sets, *it is very important that students are not forced to use the strategy suggested by the hint. In fact, students should be encouraged to find solutions to problems using as many different strategies as they can.*

Before you start—the classroom climate

There are many factors that affect the classroom climate. Among the most important are: the appropriateness of the content (not too difficult and not too easy), the evaluation practices used by the teacher, and the teacher's attitude and actions related to problem solving. Of these, the teacher's attitude and actions are most important. The appropriateness of the problem-solving experiences in this book with regard to difficulty level was discussed earlier, and some guidelines for evaluating problem-solving performance are given later. Here are some things you should do to establish a positive climate in your classroom for problem solving:

- Be enthusiastic about problem solving.
- Have students bring in problems from their personal experiences.
- Personalize problems whenever possible (e.g., use students' names).
- Recognize and reinforce willingness and perseverance.
- Reward risk takers.
- Encourage students to play hunches.
- Accept unusual solutions.
- Praise students for getting correct solutions, but during problem solving, emphasize the selection and use of problem-solving strategies.
- Emphasize persistence rather than speed.

In the first two months of the school year, the most important goal with regard to problem solving should be to establish a positive classroom climate. After this time period, you can then begin to focus on the development of the students' problem-solving abilities. The importance of a positive classroom climate cannot be overemphasized in building a successful problem-solving program.

Using the program

Although this material can be used in a variety of ways, our experience shows that it is most effectively used when problem-solving experiences are done on a regular basis and when the teacher plays an active role guiding students through the problem-solving experiences. Following are some of the ways we have successfully used the program. Each is illustrated with a scenario.

1. *Use a problem-of-the-day approach.*
The teacher starts each class with a problem-solving experience. The teacher distributes the problem to students (or has it on the chalkboard or overhead projector). The teacher then introduces the problem-solving experience, moves around the room while students are working and provides assistance when necessary, and then discusses the students' work with the class. For assisting the students and for discussing their work with them, the teacher uses the teaching notes provided. Experience has shown that the teacher will need approximately the following amounts of time for the different types of problem-solving experiences.

Problem-solving experience	*Approximate time needed for each*
Skill activities	3 to 5 minutes
One-step problems	3 to 5 minutes
Multiple-step problems	10 to 15 minutes
Process problems	15 to 20 minutes

2. *Use problem-solving experiences for homework.*
Here, the teacher gives the class a problem-solving experience for homework several nights each week. The day after a problem is assigned, the teacher discusses the problem with the class.

3. *Use a problem-of-the-week format.*
The teacher gives a multiple-step or process problem to the class on Monday and tells students that they have a week to work on it. The teacher may allow students to work together on some problems and ask the teacher for hints throughout the week. On Friday the problem is discussed in class.

4. *Have a problem-solving bulletin board.*
On Monday the teacher places a multiple-step or process problem on the problem-solving bulletin board. The students know they have the week to work on the problem and that it will be discussed in class on Friday. If they get stumped, there's an envelope on the bulletin board from which they can get hints taken from the teacher notes for solving the problem.

5. *Have a problem-solving resource box (resource center).*
The teacher has a table in the back of the room containing three boxes of cards. One box contains the multiple-step and process problems from this book. The second box contains cards, numbered to correspond to the problem cards, with hints taken from the teaching notes. The third box contains solutions to the original problems plus extensions of the original problems (also taken from the teaching notes). Students are allowed to use the resource box during any free time.

Using the blackline masters

Packaged separately from this book are 30 sheets of blackline masters. Each sheet contains one set of problem-solving experiences—one skill activity, one one-step problem, one multiple-step problem, and two process problems for a total of five experiences (see the sample provided on page xi). The last digit of the five-digit number at the bottom of each blackline master identifies its grade level. The publisher grants permission for these sheets to be duplicated on appropriate duplicating equipment. Depending upon the way you are using the program, you may want to duplicate one side of the sheet and cut it up into individual problems and pass out a single problem to the students. Or you may want to cut out each problem and paste it on tag board or laminate it, if you are using the approach of a resource box. Rather than producing a handout for each student, you may want to use a blackline master to produce a transparency for use with the entire class.

Using the teaching notes

The teaching notes for skill activities and one-step problems consist of complete solutions to the activity or problem. (Answers may vary for many skill activities, so *possible* answers are provided.) There are no specific teaching actions given in the teaching notes for skill activities and one-step problems. Most teachers simply discuss the students' work after they have completed the activity or problem. For skill activities where answers may vary, you should show as many different answers as possible. For one-step problems, students can be asked the following questions as a way to discuss their work:

1. What were you trying to find?
2. Which data in the story were needed to find the solution? Were there any unnecessary data?
3. What action in the story suggested the operation you used to find the answer?
4. Can you give the answer in a complete sentence?
5. Have you checked your work and your answer?

Most teachers find it much more difficult to teach multiple-step and process problems than to teach skill activities and one-step problems. Therefore, the teaching notes for these types of experiences are quite extensive. The teaching notes are in the form of a lesson plan (see, for example, page 2). The three circles on the plan (Teaching Actions Before, Teaching Actions During, Teaching Actions After) are three categories of teacher behaviors to be used for guiding students through their work on a problem. The table on the next page gives a complete description of each teaching action, and describes the purpose of each. A scenario is useful to illustrate how to use the teaching actions in the classroom.

> BEFORE students start work on a problem have a whole-class discussion about the problem, following teaching actions 1, 2, and 3. After this discussion, have students begin working on the problem (preferably in small groups). DURING the time they are working on the problem, move around the room monitoring and directing students' work (teaching actions 4–7). Near the end of the time students are working on the problem, have two or three students place their solutions on the chalkboard. Then, AFTER they have solved the problem, have another whole-class discussion about the students' work (teaching actions 8, 9, and 10).

The teaching notes for each multiple-step problem and process problem give much more than a general list of teaching actions. One of the key elements in successfully guiding students' problem-solving experiences is asking the right questions at the right time. For each multiple-step and process problem, classroom-tested questions and hints are given in the lesson plan. The first set of questions (*Understanding the Problem*) should be used BEFORE students start work when you are helping them understand the problem (teaching action 2). The second set of questions (*Planning a Solution)* should be used DURING the time students are working on a problem, if or when they get stumped in their solution attempt (teaching action 5). The hints given for *Planning a Solution* should be viewed as *possible* hints. As you observe and question students, you must decide which, if any, of those hints are appropriate. Sometimes, none of the hints listed will seem appropriate, and you will need to come up with others. Quite often you'll find it necessary to repeat one or more of the Understanding the Problem questions you used in the whole-class discussion BEFORE students started work. Most teachers find that selecting just the right hint for a student or group is a teaching skill that develops with experience.

Teaching Actions	Purpose of Teaching Action
Before	
1. Read the problem to the class or have a student read the problem. Discuss words or phrases students may not understand.	To illustrate the importance of reading problems carefully and to focus on words that have special interpretations in mathematics.
2. Use a whole-class discussion about understanding the problem. Ask questions to help students understand the problem (see the problem-specific hints for Understanding the Problem).	To focus attention on important data in the problem and to clarify parts of the problem.
3. Ask students which strategies might be helpful for finding a solution. Do not evaluate students' suggestions. You can direct students' attention to the list of strategies on the problem-solving guide when asking for suggestions (see page xiii).	To elicit ideas for *possible* ways to solve a problem.
During	
4. Observe and question students about their work.	To diagnose students' strengths and weaknesses related to problem solving.
5. Give hints for solving the problem as needed. (See the problem-specific hints for Planning a Solution.)	To help students get past blocks in solving a problem.
6. Require students who obtain a solution to check their work and answer the problem. You can direct students' attention to the "Checking Work and Answering the Problem" section on the problem-solving guide (page xiii) for specific ways to check and answer the problem.	To require students to look over their work.
7. Give a problem extension to students who complete the original problem much sooner than others. (See the Problem Extension section.)	To keep all students involved in a meaningful problem-solving experience until others have completed work on the original problem. (This is a classroom management teaching action. See teaching action 9 for using problem extensions to improve problem-solving ability.)
After	
8. Show and discuss students' solutions to the original problem. Have students name the strategies used. You can reinforce the names of the strategies by pointing out the strategy names on the problem-solving guide (page xiii).	To show and name strategies for solving the problem.
9. Relate the problem to previous problems (if possible) and solve an extension of the original problem. (See the Related Problems and Problem Extension sections.)	To demonstrate that problem-solving strategies are not problem-specific and to help students recognize different kinds of situations in which particular strategies may be useful.
10. Discuss special features of the original problem, if any. (See the Comment section.)	To show how special features of problems (for example, a picture accompanying the problem statement) may influence students' thinking.

This problem contains unnecessary data. Select the data you need to solve the problem. Then find the answer.

Milkshakes cost $0.80 each. Cheeseburgers cost $0.75 each. Manuel bought 2 cheeseburgers. How much did he spend for the cheeseburgers?

16

Ryan made seven foul shots out of twelve. Steve made eight foul shots out of thirteen. How many foul shots did they make in all?

17

When you are 5 km from the lake, how far are you from the city?

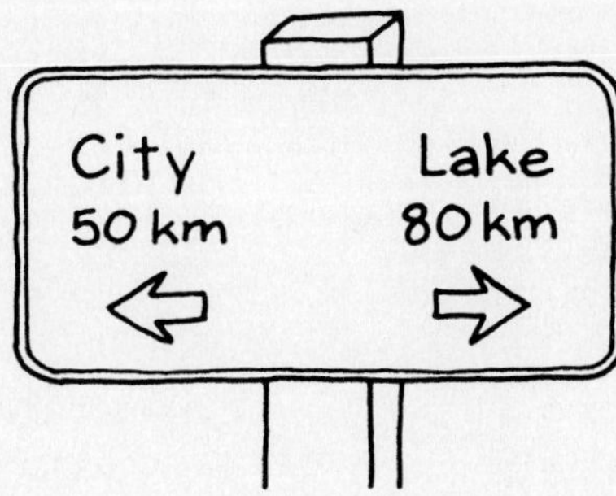

18

Holly checked a book out of the library and read this notice about fines: If a book is 1 day overdue, the fine is 1¢, 2 days overdue, 2¢, 3 days overdue, 4¢, and so on. If Holly's book is 7 days overdue, how much is her fine? (Hint: Complete this table.)

Day	1	2	3	4	5	6	7
Fine	1¢	2¢	4¢	8¢			

19

Seth and Bob each began reading a Hardy Boys book today. If Seth reads 8 pages each day and Bob reads 5 pages each day, what page will Bob be reading when Seth is reading page 56? (Hint: Complete the table.)

Day	*Seth's Page*	*Bob's Page*
1	8	5
2	16	10
3	24	15

20

In the *Finding the Answer* section, at least one solution to the problem is shown. The names of the solution strategies are given, and the answer to the problem is given in a complete sentence. The solutions shown for each problem were selected because they are ones used most often by students in our work with these problems. However, it is possible that students will use solution strategies different from the ones shown. That's fine! *Students should not be required to use a particular solution strategy for a given problem. Rather, students should be encouraged to find as many ways as possible to solve problems.*

The *Related Problems* section identifies (by number) problems that appeared earlier in the book that can be solved in ways similar to the given problem. The *Problem Extension* section includes an additional problem that is similar to the original problem. The answer to the problem extension is provided after the problem statement. Some problems have a *Comment* section containing an observation about the problem that could be used with teaching action 10 (discuss any special features of the problem). For example, some pictures accompanying problem statements can be misleading, and a statement to this effect could appear in this section.

Most teachers find the teaching actions provide a welcome structure for guiding students' problem-solving experiences yet they are flexible enough to be adapted to individual teaching styles. To further help implement the teaching actions, teachers have found the problem-solving guide shown below quite helpful.

Student Evaluation

You should have a plan for evaluating your students' problem-solving performance. A pupil evaluation plan for this program should not be limited to a check for correct answers. Instead, you should attempt to assess many of the process and attitudinal goals of the program. Since problem-solving performance improves over a relatively long period of time, it is not necessary that you evaluate each student's performance in every problem-solving situation. Rather, it is sufficient to assess students periodically, and, over a period of time (perhaps the entire year), an adequate profile of a child's performance capabilities and improvements can be obtained. We have found the following two types of evaluative data to be helpful: (1) scores from an analysis of students' written work on problems and (2) observations of students' work. A point system for analyzing students' written work on problems and an observation checklist are provided on page xv.

Solving the Problem

- Guess and check.
- Draw a picture.
- Make an organized list.
- Make a table.
- Work backwards.
- Look for a pattern.
- Use logical reasoning.
- Choose the operation(s).

Checking Work and Answering the Problem

- Be sure you used all the important information.
- Check any arithmetic.
- Decide whether the answer makes sense.
- Write the answer in a complete sentence.

Some special considerations

Success in implementing a program can often depend on the resolution of seemingly minor issues that we have found important in the implementation of the program.

1. *Time needed to implement the program.* The amount of time needed to implement PROBLEM-SOLVING EXPERIENCES obviously depends on the ways you use the program. The chart on page viii shows the time you might expect to spend on each type of problem-solving experience if you use the complete set of teaching notes during a class period. Regardless of the ways you choose to implement PROBLEM-SOLVING EXPERIENCES *it is important to realize that the time needed for each experience will be greatest at the beginning of the year.* This is particularly true if your students have not had prior experience in a problem-solving program. You may feel that too much time is being spent on problem solving at the beginning of the year, but don't stop! After a while, you'll find that the time needed to work in class on a problem decreases. If you choose a problem-of-the-day approach and if you feel that the time you're spending is excessive, you might try (1) omitting the one-step problems, or (2) assigning the one-step problems as homework or (3) assigning the skill activities as homework. These techniques reduce the amount of class time needed to implement the program, yet still enable you to provide at least three in-class problem-solving experiences each week.

2. *Grouping students for instruction.* If you choose to have students work on the experiences in class, you must decide how you will group students for instruction. We recommend that students work individually on skill activities and one-step problems. These experiences are relatively easy and can be completed very quickly by students working alone. We recommend small-group work for multiple-step and process problems (no more than four students per group). Since multiple-step and process problems are usually challenging, small-group work helps reduce the pressure on the individual student. A second reason to use small groups is that it's easier to monitor and assess the problem-solving performance of all the students in the class. Third, small-group work often elicits behaviors that promotes the improvement of problem-solving performance such as justifying and evaluating ideas.

3. *Calculators. None of the problem-solving experiences in this program requires the use of a calculator.* However, there are several situations in which you might consider allowing students to use a calculator. Problems involving the guess-and-check strategy are good candidates for allowing the use of a calculator because calculators minimize the time required to check a guess. *Pattern-finding* problems are also appropriate for calculators because numerical patterns can be tested rapidly. Problems involving the *work backwards* strategy and problems involving the *choose-the-operations strategy* require several computational steps to find a solution. Using calculators for these problems allows students to focus on how the action in the story suggests the operations needed to find a solution rather than on the computation. Finally, since most problem-solving experiences involve some computation, *the calculator may be the only vehicle for students who are poor at computation to participate in problem solving.* We recommend the availability of a calculator for these students.

4. *Reorganizing the problem-solving experiences.* Should you decide to use only some of the problem-solving experiences in this book, the organization of the experiences has implications for how you should select and use them.

a. *Do not start in the middle of the book.* For each type of problem-solving experience, we have attempted to sequence the problems from easy to difficult. For example, the first multiple-step problem will probably be easier for students than the next one, and so on through the 30 multiple-step problems. This easy-to-difficult sequencing of experiences, as well as the way hints are introduced for process problems and then omitted, makes it inappropriate to start using problem-solving experiences at any place in the book. If you do not start the program at the beginning of the school year, you should still begin with the first problem-solving experience in the book and move sequentially through the program.

b. *Do not compact the one-step and multiple-step problems.* The computational skills required to solve the one-step and multiple-step problems in PROBLEM-SOLVING EXPERIENCES are ones students will most likely have already been taught in their regular mathematics program if you begin using this program at the beginning of the school year and you use a problem-of-the-day approach. Because of the organization of these problems in relation to the content of the textbooks, it is inappropriate to use all of the one-step and multiple-step problems, for example, in the first half of the school year.

A Point System for Scoring Written Work

Understanding the problem

0 Completely misinterprets the problem.

1 Misinterprets part of the problem.

2 Complete understanding of the problem.

Choosing and implementing a solution strategy

0 Makes no attempt or uses a totally inappropriate strategy.

1 Chooses a partly correct strategy based on interpreting part of the problem correctly.

2 Chooses a strategy that could lead to a correct solution if used without error.

Getting the answer

0 Gets no answer or a wrong answer based on an inappropriate solution strategy.

1 Makes copying error or computational error, gets partial answer for a problem with multiple answers, or labels answer incorrectly.

2 Gets correct solution.

Problem-Solving Observation Checklist

Student ____________________

Date ____________________

	Frequently	*Sometimes*	*Never*
1. Selects appropriate solution strategies.	______	______	______
2. Accurately implements solution strategies.	______	______	______
3. Tries a different solution strategy when stuck (without help from the teacher).	______	______	______
4. Approaches problems in a systematic manner (clarifies the question, identifies needed data, selects and implements a solution strategy, checks solution).	______	______	______
5. Shows a willingness to try problems.	______	______	______
6. Demonstrates self-confidence.	______	______	______

1 Skill Activity

Write a question that can be answered by using the data in this story. Then find the answer.

George weighed 140 pounds (lb). He went on a diet and lost 15 lb.

Discussion

Possible question:

How much does George weigh now? (125 lb)

2 One-Step Problem

Linda is 9 years old. Her brother, Joe, is 4 years older than she. How old is Joe?

Solution

$$\begin{array}{r} 9 \\ +\ 4 \\ \hline 13 \end{array}$$

Joe is 13 years old.

3 Multiple-Step Problem

Ms. Baxter was 46 years old in 1984. How old was she in 1970?

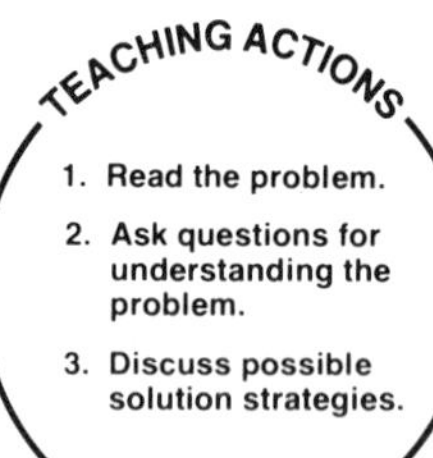

Understanding the Problem

- How old was Ms. Baxter in 1984? (46)
- What are we trying to find? (her age in 1970)

TEACHING ACTIONS

4. Observe students.
5. Give hints as needed for solving the problem.
6. Require students to check back and answer the problem.
7. Give problem extension as needed.

DURING

Planning a Solution

- How many years are there between 1970 and 1984? (14)
- Was she younger or older in 1970? (younger)
- If you know how many years it is between 1970 and 1984, which operation would you use to find out how old Ms. Baxter was in 1970? (subtraction)

TEACHING ACTIONS

8. Discuss solution(s).
9. Discuss related problems and extension.
10. Discuss special features as needed.

AFTER

Finding the Answer

Choose the Operations

$$\begin{array}{r} 1984 \\ -\underline{1970} \\ 14 \end{array} \rightarrow \begin{array}{r} 46 \\ -\underline{14} \\ 32 \end{array}$$

Ms. Baxter was 32 years old in 1970.

Problem Extension

If Ms. Baxter was twice as old as her son in 1984, how old was he in 1970? (9)

4 Process Problem

Mr. Singer told the children to open their math books to the facing pages whose page numbers add up to 85. To which pages should the children turn? (Hint: Make a guess. Then check your guess.)

TEACHING ACTIONS

1. Read the problem.
2. Ask questions for understanding the problem.
3. Discuss possible solution strategies.

BEFORE

Understanding the Problem

- What is the sum of the page numbers? (85)
- Could the pages be 20 and 65? (No, they have to be consecutive.)
- Suppose the sum of the page numbers was 5. What would the page numbers be? $(2 + 3 = 5)$

TEACHING ACTIONS

4. Observe students.
5. Give hints as needed for solving the problem.
6. Require students to check back and answer the problem.
7. Give problem extension as needed.

DURING

Planning a Solution

- What page comes before 20? (19)
- Could these be the correct facing pages? (No, because $20 + 19 \neq 85$.)
- Could the pages be 20 and 21? (No, their sum is not 85.)
- Choose two consecutive page numbers and find their sum.

TEACHING ACTIONS

8. Discuss solution(s).
9. Discuss related problems and extension.
10. Discuss special features as needed.

AFTER

Finding the Answer

Guess and Check

- Try 30,31—$30 + 31 = 61$ (too low)
- Try 40,41—$40 + 41 = 81$ (too low)
- Try 42,43—$42 + 43 = 85$ (correct)

The children should open their math books to pages 42 and 43.

Problem Extension

Terri accidentally tore out a page of her math book. Mr. Singer asked her what page it was. Terri said that the sum of the page numbers on the facing pages was 127. What are the page numbers on the pages Terri tore out of her book? (63 and 64)

5 Process Problem

I wrote 5 different numbers on 5 cards. The sum of the numbers is 15. What numbers did I put on the cards? (Hint: Make a guess. Then check your guess.)

TEACHING ACTIONS — BEFORE

1. Read the problem.
2. Ask questions for understanding the problem.
3. Discuss possible solution strategies.

Understanding the Problem

- How many numbers did I write? (5)
- What is the sum of the numbers? (15)
- How many numbers on each card? (1)
- Are any 2 numbers the same? (No, they are all different.)

TEACHING ACTIONS — DURING

4. Observe students.
5. Give hints as needed for solving the problem.
6. Require students to check back and answer the problem.
7. Give problem extension as needed.

Planning a Solution

- Could 1 of the numbers be 15? (No, because the rest would be 0 and we said that all the numbers were different.)
- Select 5 numbers and check to see if their sum is 15.

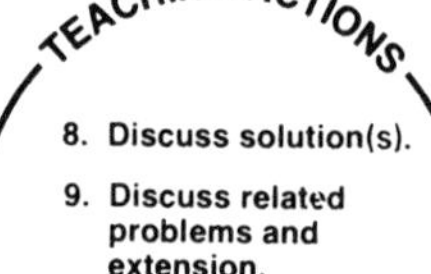

TEACHING ACTIONS — AFTER

8. Discuss solution(s).
9. Discuss related problems and extension.
10. Discuss special features as needed.

Finding the Answer

Guess and Check

- Try 0,1,2,3,4—0 + 1 + 2 + 3 + 4 = 10 (too little)
- Try 1,2,3,4,5—1 + 2 + 3 + 4 + 5 = 15 (correct)

The numbers are 1, 2, 3, 4, and 5.

Related Problem: 4

Problem Extension

If the numbers are even and different and their sum is 30, what are the numbers? (2,4,6,8,10)

6 Skill Activity

Write a question that can be answered by using the data in this story. Then find the answer.

Jean had \$12.50 last week. On Saturday she received \$3.00 for her birthday and \$2.25 for babysitting.

Discussion

Possible question:

How much does Jean have now? (\$17.75)

7 One-Step Problem

I went to town with \$14.87. When I returned, I had \$6.39. How much did I spend?

Solution

$$\begin{array}{r} \$14.87 \\ -\ 6.39 \\ \hline \$\ 8.48 \end{array}$$

I spent \$8.48 in town.

8 Multiple-Step Problem

Mr. and Mrs. Mason and their 3 children paid $7.00 for tickets. Which game do they plan to see?

Softball	Basketball	Soccer
Adults $2.00	Adults $3.00	Adults $5.00
Children $1.00	Children $1.50	Children $2.50

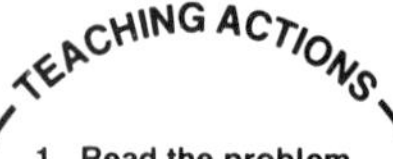

1. Read the problem.
2. Ask questions for understanding the problem.
3. Discuss possible solution strategies.

BEFORE

Understanding the Problem

- How many adults are there? (2)
- How many children are there? (3)
- How much did they spend? ($7.00)
- Do adults' and children's prices mean "each"? (yes)

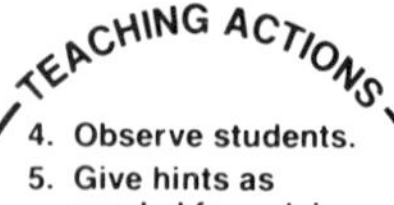

4. Observe students.
5. Give hints as needed for solving the problem.
6. Require students to check back and answer the problem.
7. Give problem extension as needed.

DURING

Planning a Solution

- How much do 2 adults pay for soccer? ($10.00)
- Could the Mason family go to the soccer game? (No, they only spent $7.00 in all.)
- How much do 2 adults pay for softball? ($4.00)

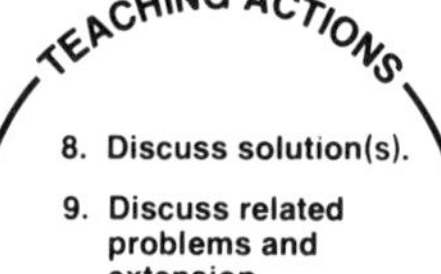

8. Discuss solution(s).
9. Discuss related problems and extension.
10. Discuss special features as needed.

AFTER

Finding the Answer

Choose the Operations

$2.00		$1.00		$4.00
× 2	→	× 3	→	+ 3.00
$4.00		$3.00		$7.00

They plan to go to the softball game.

Problem Extension

How much change would Mr. and Mrs. Mason get if they took their children to see the basketball game and paid for it with a $20.00 bill? ($9.50)

9 Process Problem

Janet and Vicki put up a rope to mark the starting line for the sack race. The rope was 10 meters (m) long. They put a post at each end of the rope and at every 2 m. How many posts did they use? (Hint: Finish drawing the picture to help you.)

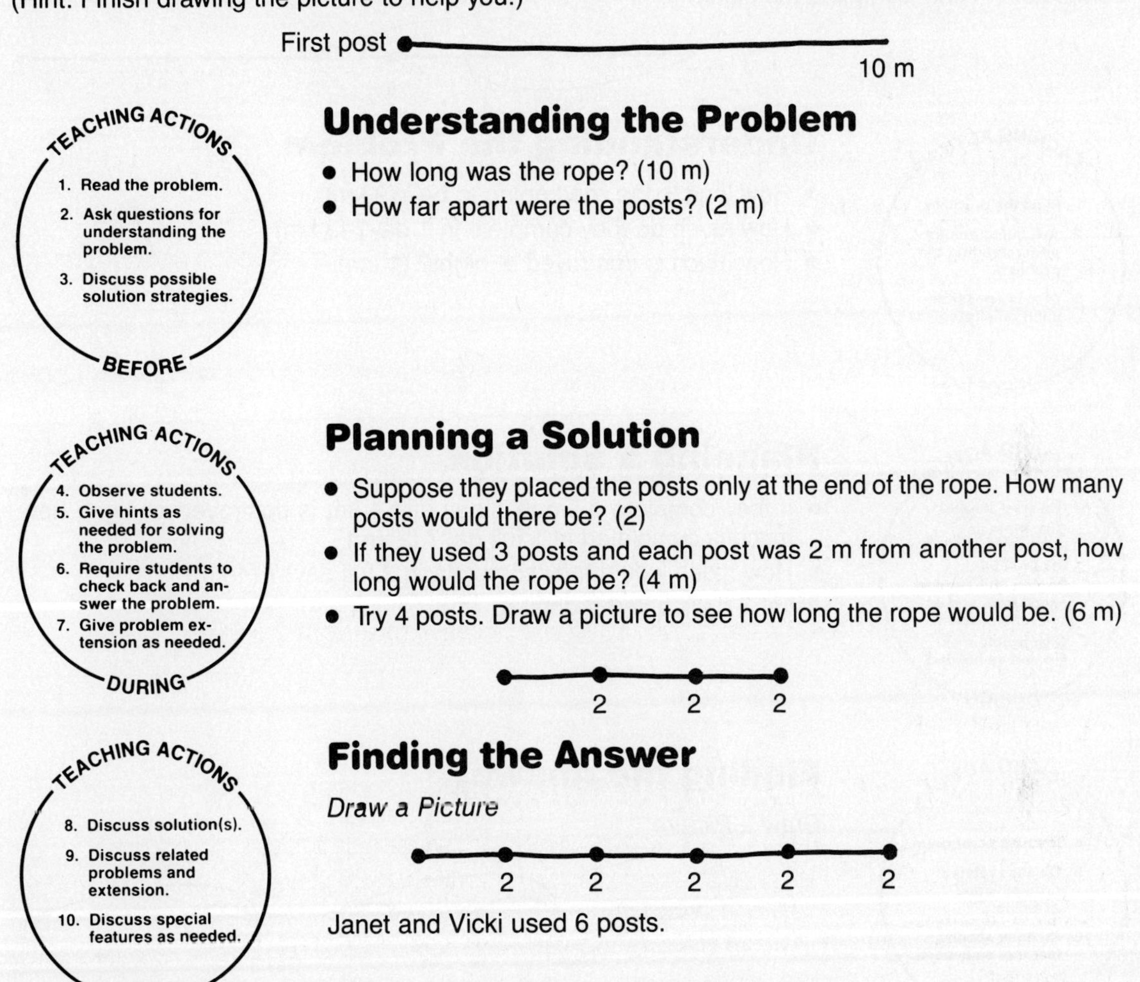

Understanding the Problem

- How long was the rope? (10 m)
- How far apart were the posts? (2 m)

Planning a Solution

- Suppose they placed the posts only at the end of the rope. How many posts would there be? (2)
- If they used 3 posts and each post was 2 m from another post, how long would the rope be? (4 m)
- Try 4 posts. Draw a picture to see how long the rope would be. (6 m)

2 2 2

Finding the Answer

Draw a Picture

2 2 2 2 2

Janet and Vicki used 6 posts.

Problem Extension

Mr. Brown put a square fence around his vegetable garden to keep the deer from eating his corn. Each side was 10 m. If the posts were placed 2 m apart, how many posts did he use? (20)

10 Process Problem

A road crew is building a 9-kilometer (km) road along the side of a mountain. Each day they complete 3 km, but each night rock slides destroy 1 km of the road. At this rate, on what day will the road be completed? (Hint: Complete the picture.)

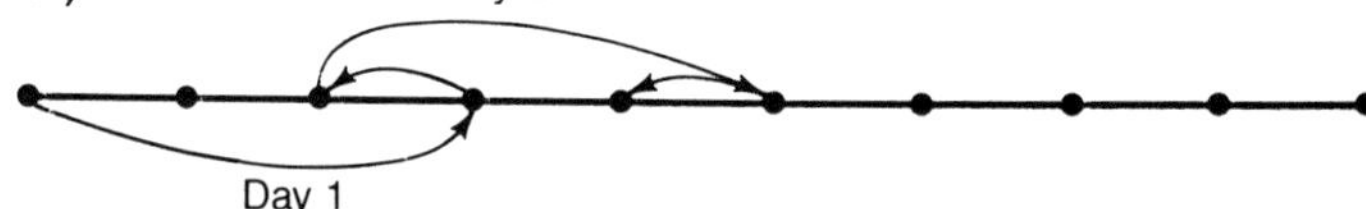

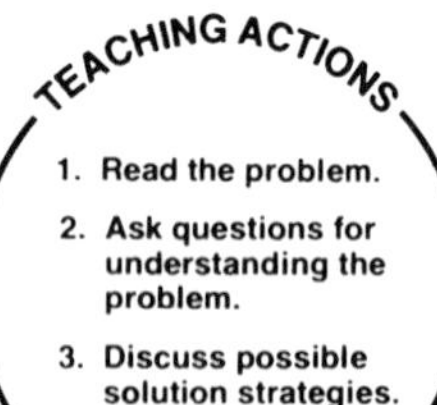

Understanding the Problem

- How long is the road going to be? (9 km)
- How much do they complete in 1 day? (3 km)
- How much is destroyed at night? (1 km)

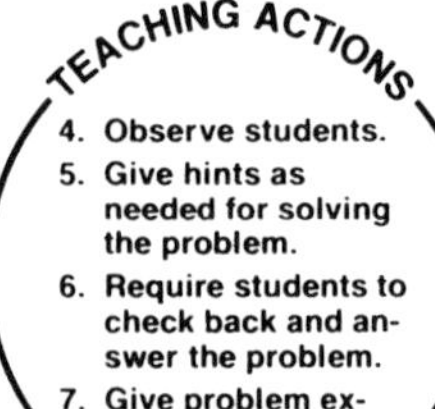

Planning a Solution

- If they complete 3 km in 1 day and 1 km is destroyed at night, how much is completed in 1 full day? (2 km)
- How much would they complete in 2 full days? (4 km)
- Try 3 days and 3 nights. (6 km)

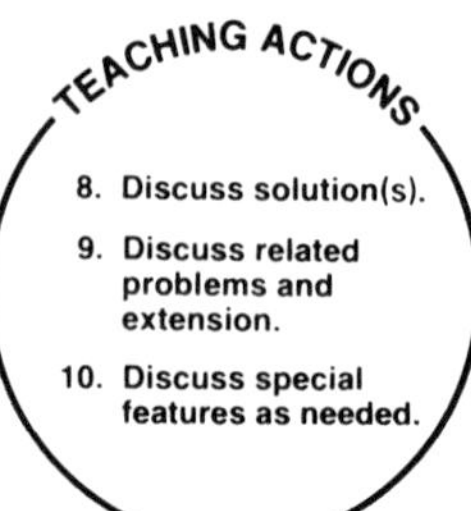

Finding the Answer

Draw a Picture

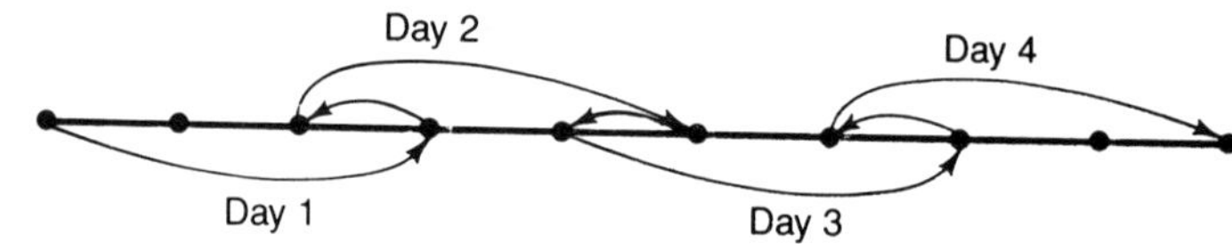

The road will be completed on the fourth day.

Related Problem: 9

Problem Extension

Suppose the crew was building a road 16 km long and they completed 4 km per day. 2 km per night were destroyed. On what day would they complete this road? (on the seventh day)

11 Skill Activity

Here are two problems and their solutions. Write the answer to each in a complete sentence.

1. Marie had 596 stamps. She bought 117 stamps. How many stamps did she have then? Answer: 713.

2. There are 10 apples on the table. Jesse ate 3, Florrie ate 2, and Rachel ate 4. How many apples were left? Answer: 1

Discussion

Possible answers:

1. Marie then had 713 stamps.

2. There was 1 apple left.

12 One-Step Problem

Valerie bought a sandwich and milk. How much did she spend for her lunch?

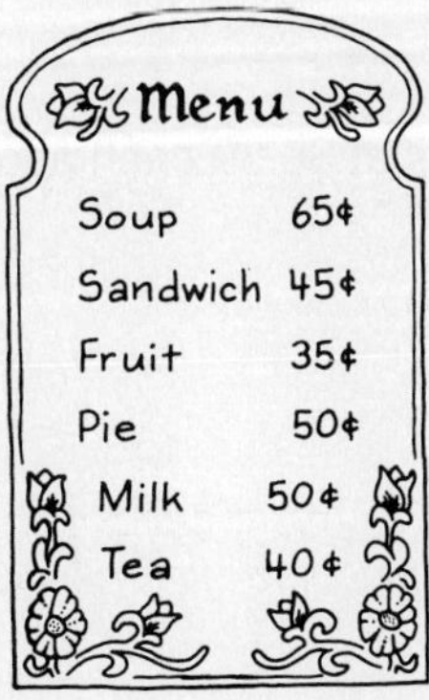

Solution

$$\begin{array}{r} 45¢ \\ +\ 50¢ \\ \hline 95¢ \end{array}$$

Valerie spent 95¢ for her lunch.

13 Multiple-Step Problem

Luz bought a sandwich for $0.39 and an apple drink for $0.55. If she paid for it with a $1.00 bill, what was her change?

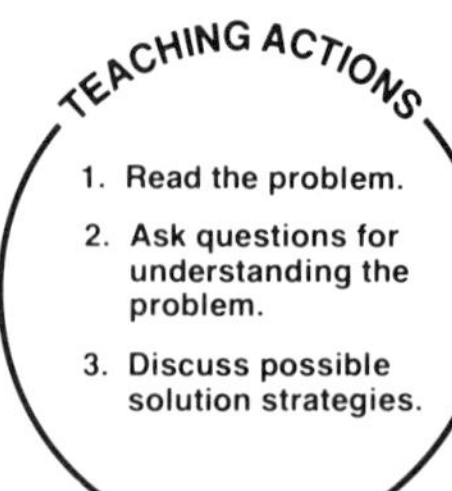

Understanding the Problem

- How much was the sandwich? ($0.39)
- How much was the apple drink? ($0.55)
- How much money did she give to pay for her lunch? ($1.00)

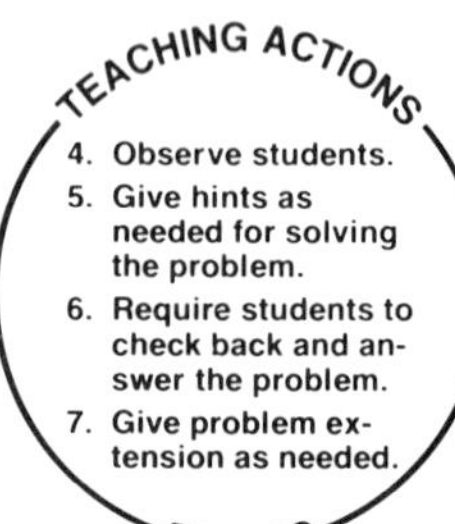

Planning a Solution

- How much did her lunch cost? ($0.94)
- If you know the cost of her lunch, which operation would you use to find out how much change she got back? (subtraction)

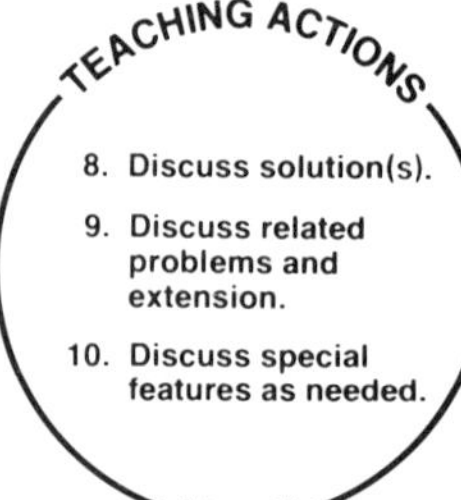

Finding the Answer

Choose the Operations

$$\begin{array}{r} \$0.39 \\ +\ 0.55 \\ \hline \$0.94 \end{array} \rightarrow \begin{array}{r} \$1.00 \\ -\ 0.94 \\ \hline \$0.06 \end{array}$$

Luz's change was $0.06.

Problem Extension

Luz only has $1.00. If she wants 2 sandwiches at $0.39 each and an apple drink, how much more money will she need? ($0.33)

14 Process Problem

Allison had lunch with her mother. She decided to order 1 sandwich and 1 drink from the menu. How many different lunches can Allison order from this menu? (Hint: Complete the organized list.)

Roast beef—Tea
Roast beef—Milk
Roast beef—Juice

BLT—Tea
BLT—Milk
BLT—Juice

Lunch Menu

Sandwiches:
Roast beef
Bacon, lettuce, and tomato
Hamburger

Drinks: Tea
Milk
Juice

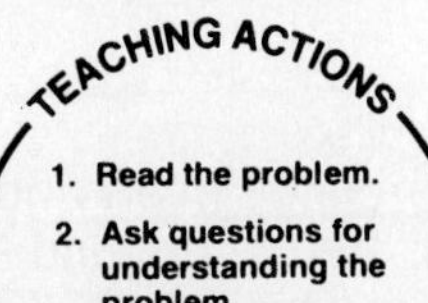

Understanding the Problem

- How many choices does Allison have for sandwiches? (3) For drinks? (3)
- What is a lunch? (sandwich and drink)
- How many sandwiches may Allison order? (1) Drinks? (1)

TEACHING ACTIONS
4. Observe students.
5. Give hints as needed for solving the problem.
6. Require students to check back and answer the problem.
7. Give problem extension as needed.
DURING

Planning a Solution

- If Allison orders a roast beef sandwich, what drinks can she order? (tea, milk, juice) So, how many different lunches can she order if she orders a roast beef sandwich? (3)
- List the lunches Allison can order if she chooses a bacon, lettuce and tomato sandwich. (BLT—tea, BLT—milk, BLT—juice)

Finding the Answer

Make an Organized List

Roast beef—Tea
Roast beef—Milk
Roast beef—Juice

BLT—Tea
BLT—Milk
BLT—Juice

Hamburger—Tea
Hamburger—Milk
Hamburger—Juice

Problem Extension

For dessert, Allison wanted an ice-cream cone. She had a choice of 4 different flavors: vanilla, strawberry, chocolate, and peach. Allison ordered a double dip. How many choices does she have for a double-dip cone? (10)

15 Process Problem

Bob had 2 tickets for rides at the carnival. There were 4 rides he could take: merry-go-round, ferris wheel, airplanes, and roller coaster. How many different ways could Bob use his 2 tickets? (Hint: Complete the organized list.)

Merry-go-round—Ferris wheel
Merry-go-round—Airplanes
Merry-go-round—Roller coaster

Ferris wheel—

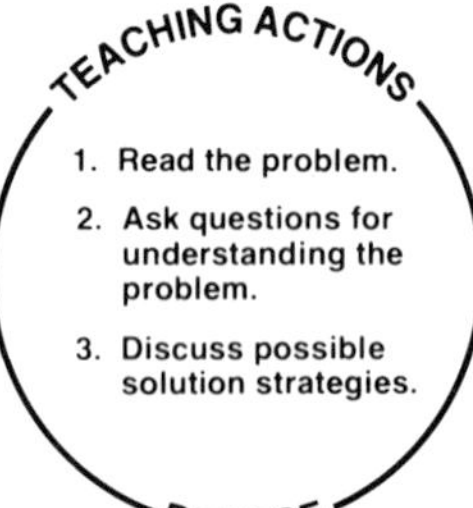

Understanding the Problem

- How many tickets does he have? (2)
- How many different rides are there? (4)

TEACHING ACTIONS

4. Observe students.
5. Give hints as needed for solving the problem.
6. Require students to check back and answer the problem.
7. Give problem extension as needed.

DURING

Planning a Solution

- What else could he ride if he rode the merry-go-round? (merry-go-round—ferris wheel, merry-go-round—airplanes, merry-go-round—roller coaster) How many different ways has he used his tickets? (3)
- If one way Bob uses his tickets is to ride the merry-go-round and the ferris wheel, would a second way be to ride the ferris wheel and then the merry-go-round? (No, the order doesn't matter.)

Finding the Answer

Make an Organized List

Merry-go-round—Ferris wheel
Merry-go-round—Airplanes
Merry-go-round—Roller coaster

Ferris wheel—Airplanes
Ferris wheel—Roller coaster

Airplanes—Roller coaster

Bob could use his tickets 6 different ways.

Related Problem: 14

Problem Extension

If Bob had 3 tickets, how many different ways could he ride the 4 rides? (4)

16 Skill Activity

This problem contains unnecessary data. Select the data you need to solve the problem. Then find the answer.

Milkshakes cost $0.80 each. Cheeseburgers cost $0.75 each. Manuel bought 2 cheeseburgers. How much did he spend for the cheeseburgers?

Discussion

You don't need to know the cost of the milkshakes.

$$\begin{array}{r} \$0.75 \\ +\ 0.75 \\ \hline \$1.50 \end{array} \quad \textit{or} \quad \begin{array}{r} \$0.75 \\ \times\ 2 \\ \hline \$1.50 \end{array}$$

Manuel spent $1.50 for 2 cheeseburgers.

17 One-Step Problem

Ryan made 7 foul shots out of 12. Steve made 8 foul shots out of 13. How many foul shots did they make in all?

Solution

$$\begin{array}{r} 7 \\ +\ 8 \\ \hline 15 \end{array}$$

Ryan and Steve made 15 foul shots in all.

18 Multiple-Step Problem

When you are 5 km from the lake, how far are you from the city?

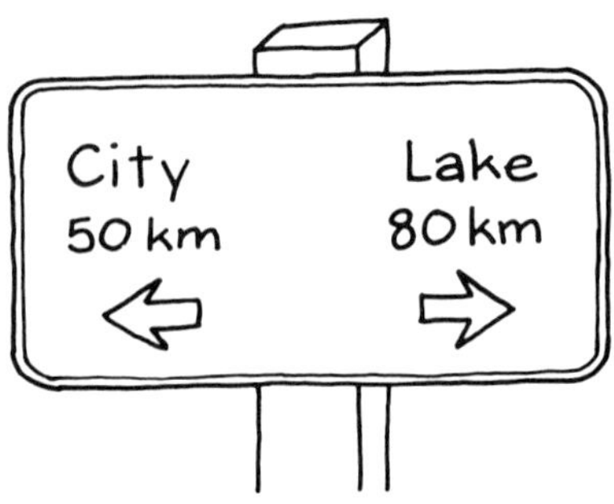

TEACHING ACTIONS

1. Read the problem.
2. Ask questions for understanding the problem.
3. Discuss possible solution strategies.

BEFORE

Understanding the Problem

- How far are you from the lake? (5 km)

TEACHING ACTIONS

4. Observe students.
5. Give hints as needed for solving the problem.
6. Require students to check back and answer the problem.
7. Give problem extension as needed.

DURING

Planning a Solution

- How far is it from the pole to the city? (50 km)
- How far are you from the pole? (80 km − 5 km = 75 km)

TEACHING ACTIONS

8. Discuss solution(s).
9. Discuss related problems and extension.
10. Discuss special features as needed.

AFTER

Finding the Answer

Choose the Operations

$$\begin{array}{r} 80 \\ -5 \\ \hline 75 \end{array} \rightarrow \begin{array}{r} 75 \\ +50 \\ \hline 125 \end{array}$$

When you are 5 km from the lake, you are 125 km from the city.

Related Problem: 13

Problem Extension

How long would it take you to get from the city to the lake if you rode your bicycle at a speed of 10 km per hour (km/h)? (13 h)

19 Process Problem

Holly checked a book out of the library and read this notice about fines: If a book is 1 day overdue, the fine is 1¢, 2 days overdue, 2¢, 3 days overdue, 4¢, and so on. If Holly's book is 7 days overdue, how much is her fine? (Hint: Complete this table.)

Day	1	2	3	4	5	6	7
Fine	1¢	2¢	4¢	8¢			

TEACHING ACTIONS

1. Read the problem.
2. Ask questions for understanding the problem.
3. Discuss possible solution strategies.

BEFORE

Understanding the Problem

- How much is the fine for 1 day? (1¢)
- How much is the fine for 2 days? (2¢)
- How much is the fine for 3 days? (4¢)

TEACHING ACTIONS

4. Observe students.
5. Give hints as needed for solving the problem.
6. Require students to check back and answer the problem.
7. Give problem extension as needed.

DURING

Planning a Solution

- How much would the fine be for 4 days if we double the previous day? (8¢)
- How much is the fine for 5 days? (16¢)

TEACHING ACTIONS

8. Discuss solution(s).
9. Discuss related problems and extension.
10. Discuss special features as needed.

AFTER

Finding the Answer

Make a Table

Day	1	2	3	4	5	6	7
Fine	1¢	2¢	4¢	8¢	16¢	32¢	64¢

Holly's fine is 64¢.

Problem Extension

Holly had 2 books overdue. One book was 10 days overdue and the other was 5 days overdue. What was her total fine? ($5.28)

20 Process Problem

Seth and Bob each began reading a Hardy Boys book today. If Seth reads 8 pages each day and Bob reads 5 pages each day, what page will Bob be reading when Seth is reading page 56? (Hint: Complete the table.)

Day	Seth's Page	Bob's Page
1	8	5
2	16	10
3	24	15

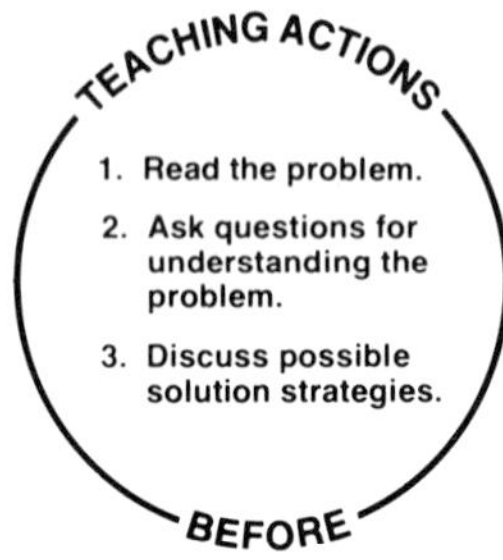

Understanding the Problem

- How many pages does Seth read each day? (8) Bob? (5)
- Did they start reading their books on the same day? (yes)

TEACHING ACTIONS

4. Observe students.
5. Give hints as needed for solving the problem.
6. Require students to check back and answer the problem.
7. Give problem extension as needed.

DURING

Planning a Solution

- How many pages had Seth read at the end of the first day? (8) Bob? (5)
- When Seth has read 16 pages, how many pages will Bob have read? (10)
- Find the number of pages Seth read for the first 5 days. (8, 16, 24, 32, 40)

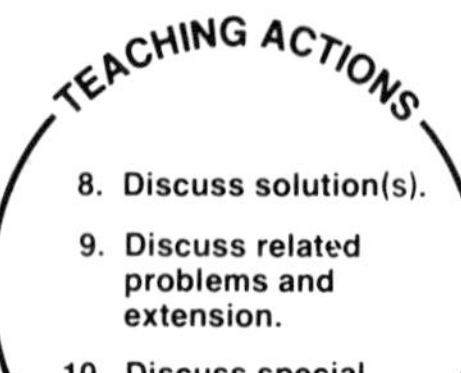

Finding the Answer

Make a Table

Day	Seth's Page	Bob's Page
1	8	5
2	16	10
3	24	15
4	32	20
5	40	25
6	48	30
7	56	35

Bob will be reading page 35 when Seth is reading page 56.

Related Problem: 19

Problem Extension

Mary reads 9 pages a day, Sue reads 10 pages a day, and Molly reads 8 pages a day. What page will Sue and Molly each be reading when Mary is reading page 72? (Sue—page 80, Molly—page 64)

21 Skill Activity

This problem has data missing. Make up appropriate data. Then solve the problem using your data.

Cindy and Spencer have $15.00 together. How much of the money is Spencer's?

Discussion

Possible answer:

Cindy's share is $7.35. (Spencer's share is $7.65.)

22 One-Step Problem

Alice had 11 rabbits and Will had 7 rabbits. Alice bought 3 more rabbits. How many rabbits does Alice have now?

Solution

$$\begin{array}{r} 11 \\ +\ 3 \\ \hline 14 \end{array}$$

Alice has 14 rabbits.

Note: The number of rabbits Will has is unnecessary data.

23 Multiple-Step Problem

Mr. Gomez left York and traveled to Freemont and then to Clay on a business trip. When he left Clay he returned to York through Ross. How much shorter was his return trip?

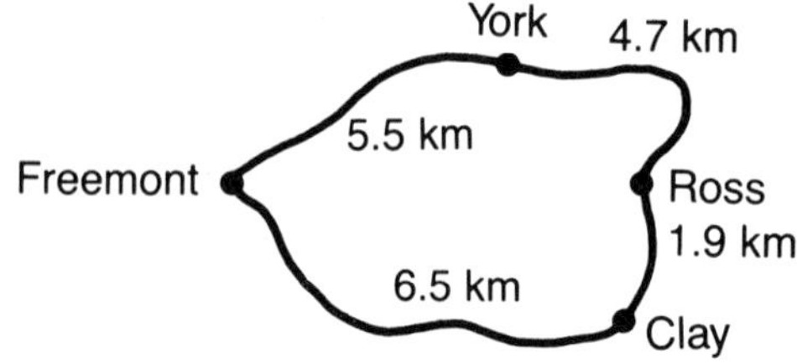

TEACHING ACTIONS

1. Read the problem.
2. Ask questions for understanding the problem.
3. Discuss possible solution strategies.

BEFORE

Understanding the Problem

- Where did Mr. Gomez go when he left York? (Freemont)
- When he left Freemont, where did he go? (Clay)
- What city did he go through when he returned? (Ross)

TEACHING ACTIONS

4. Observe students.
5. Give hints as needed for solving the problem.
6. Require students to check back and answer the problem.
7. Give problem extension as needed.

DURING

Planning a Solution

- What is the distance from York to Clay, going through Freemont? (5.5 km + 6.5 km = 12 km)
- What is the distance from Clay to York going through Ross? (1.9 km + 4.7 km = 6.6 km)

TEACHING ACTIONS

8. Discuss solution(s).
9. Discuss related problems and extension.
10. Discuss special features as needed.

AFTER

Finding the Answer

Choose the Operations

$$\begin{array}{r} 5.5 \\ +\ 6.5 \\ \hline 12.0 \end{array} \rightarrow \begin{array}{r} 1.9 \\ +4.7 \\ \hline 6.6 \end{array} \rightarrow \begin{array}{r} 12.0 \\ -\ 6.6 \\ \hline 5.4 \end{array}$$

Mr. Gomez's return trip was 5.4 km shorter.

Related Problems: 18, 13

Problem Extension

Mr. Gomez rode his bicycle on the trip, so he only traveled at about 3.1 km/h. About how long would it take him to make the complete journey from York to Freemont to Clay to Ross and back to York if he did not stop? (about 6 h)

24 Process Problem

Phil was given his allowance on Monday. On Tuesday he spent $1.50 at the fruit stand. On Wednesday, Jed paid Phil the $1.00 he owed him. If Phil now has $2.00, how much is his allowance? (Hint: Using the facts given, start with the amount Phil has now and work backwards.)

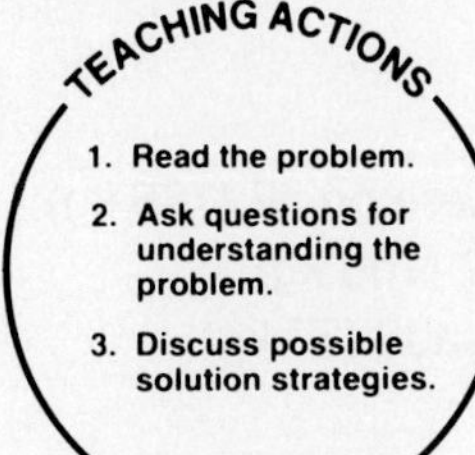

Understanding the Problem

- How much money did Phil have after Wednesday? ($2.00)
- Do you know how much Phil's allowance is? (no)
- How much did Phil spend at the fruit stand? ($1.50)
- Was Phil given any money after he got his allowance? (Yes, Jed gave him $1.00.)

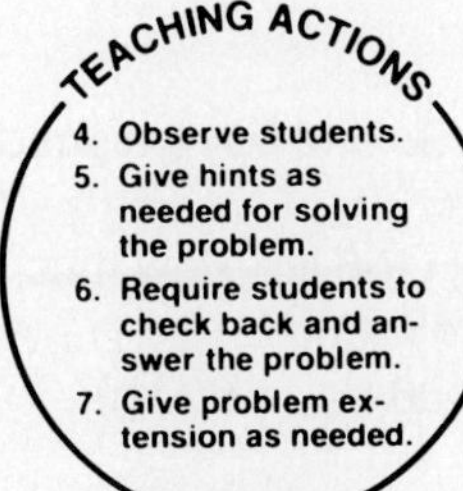

Planning a Solution

- Did Phil have Jed's $1.00 on Tuesday night? (no)
- How much money did Phil have at the end of Tuesday? ($2.00 − $1.00 = $1.00)
- Did Phil spend money on Tuesday? (yes, $1.50) How much money did Phil have before he spent the $1.50? ($1.00 + $1.50 = $2.50)

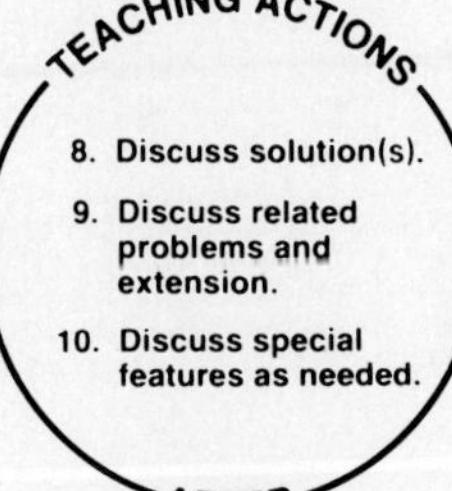

Finding the Answer

Work Backwards

↓		↑
Start with ?		End with $2.50.
Subtract $1.50.		Add $1.50.
Add $1.00.		Subtract $1.00.
End with $2.00.	So→	Start with $2.00.

Phil's allowance is $2.50.

Problem Extension

On Phil's birthday his father increased his allowance. Phil was so happy he went to the store and bought 2 cans of spray paint for his model airplanes. The paint cost $1.50. After Phil bought the paint, he had $3.50 left. How much of an increase did Phil get in his allowance? ($2.50, twice as much)

25 Process Problem

When 3 boys got on a scale, it showed 166 kilograms (kg) as their total weight. One of the boys got off and it showed 106 kg. One more boy stepped off and the scale showed 57 kg. What was the weight of each boy? (Hint: Using the facts given, start with 166 kg and work backwards.)

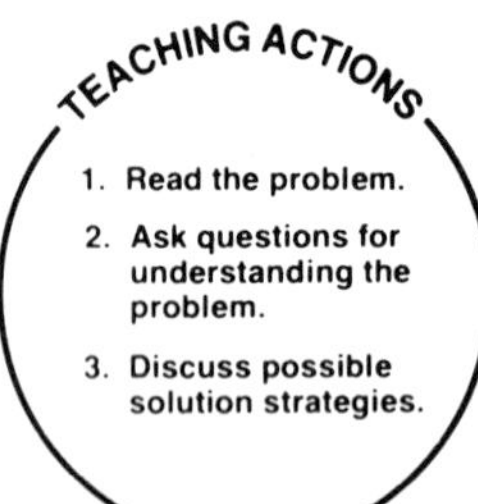

Understanding the Problem

- How many boys were on the scale at first? (3)
- How much did the scale read when all 3 boys were on it? (166 kg)
- When the first boy stepped off, what did it read? (106 kg)
- When the second boy stepped off, what did it read? (57 kg)

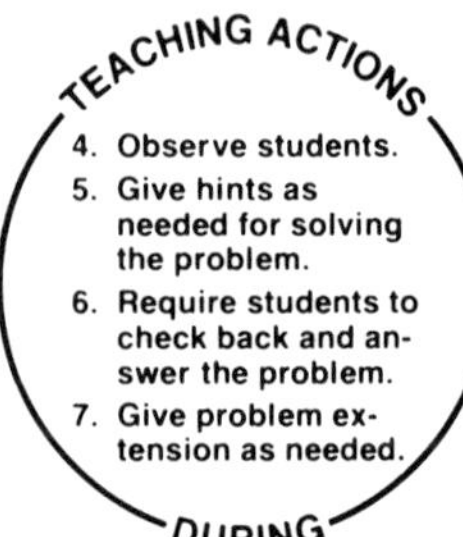

Planning a Solution

- What did the scale show with 1 boy on it? (57 kg) So one boy weighed how much? (57 kg)
- With 2 boys on it, how much did the scale show? (106 kg) How much did the second boy weigh? (106 kg − 57 kg = 49 kg)
- How much did the third boy weigh? (166 kg − 106 kg = 60 kg)

Finding the Answer

Work Backwards

Start with 166 kg.

Subtract 106—one boy weighed 60 kg.

Subtract 57—second boy weighed 49 kg.

End with 57—third boy weighed 57 kg.

The weights of the boys were 60 kg, 49 kg, and 57 kg.

Related Problem: 24

Problem Extension

There are 40 people in a restaurant. 20 drink mineral water and 14 eat spinach. 16 do neither. How many people both drink mineral water and eat spinach? (10)

26 Skill Activity

Identify the operation or operations needed to solve each problem. Then choose one problem and solve it.

1. The class set 1,250 aluminum cans as their goal in their can collection project. So far they have collected 735 cans. How many more cans do they need to meet their goal?

2. Grace, Leah, and Patty went to the beach last summer. Each of them collected 27 seashells. How many did they collect in all?

Discussion

Possible answers:

1. Use subtraction to solve this problem.

$$\begin{array}{r} 1{,}250 \\ -\ \ 735 \\ \hline 515 \end{array}$$

They need 515 cans to meet their goals.

2. Use addition or multiplication to solve this problem.

$$\begin{array}{r} 27 \\ 27 \\ +27 \\ \hline 81 \end{array} \quad \text{or} \quad \begin{array}{r} 27 \\ \times 3 \\ \hline 81 \end{array}$$

They collected 81 seashells in all.

27 One-Step Problem

Mr. Bernstein placed 210 bottles on a shelf. There were cracks in 33 of the bottles. How many were not cracked?

Solution

$$\begin{array}{r} 210 \\ -\ \ 33 \\ \hline 177 \end{array}$$

177 bottles were not cracked.

28 Multiple-Step Problem

At each of the first 2 soccer games there were 1,200 people. Last night only 900 people came to the game. Altogether, how many people have come to the 3 games?

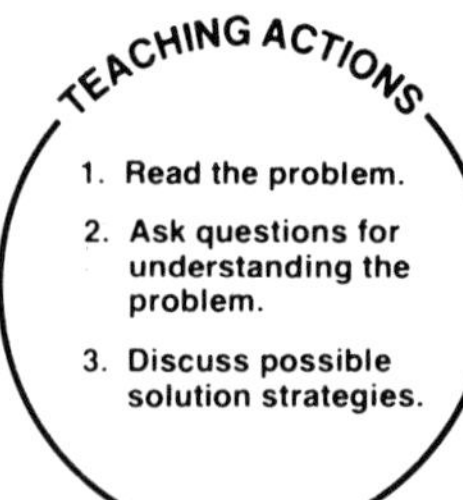

Understanding the Problem

- How many people came to each of the first 2 games? (1,200 at each game)
- How many people came to the last game? (900)

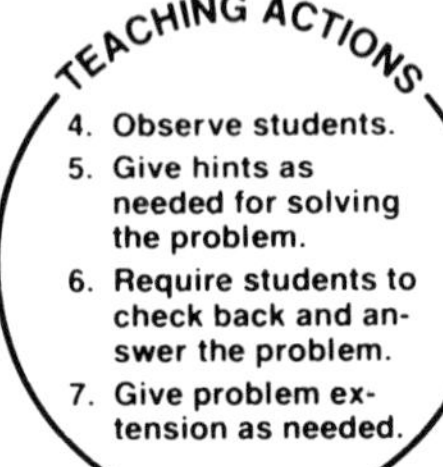

Planning a Solution

- What was the total number attending the first 2 games? ($1{,}200 + 1{,}200 = 2{,}400$)
- How many people came to the last game? (900)
- Which operation would you use to find out how many people came altogether? (addition)

Finding the Answer

Choose the Operations

$$\begin{array}{r} 1{,}200 \\ +1{,}200 \\ \hline 2{,}400 \end{array} \rightarrow \begin{array}{r} 2{,}400 \\ +\ 900 \\ \hline 3{,}300 \end{array}$$

Altogether, 3,300 people have come to the 3 games.

Problem Extension

There are 10 games in the whole season. If 1,100 people came to each of the next 3 games, 975 attended the next 2 games, and a record crowd of 1,500 attended each of the last 2 games, how many people attended the whole season? (11,550)

29 Process Problem

Ben was making trains with his new blocks. They looked like this:

First train Second train Third train

If he continues to build trains this way, how many blocks will he use in the seventh train? (Hint: Look for a pattern. Then complete the table.)

Train	Number of Blocks
1	1
2	7
3	13
4	
5	
6	
7	

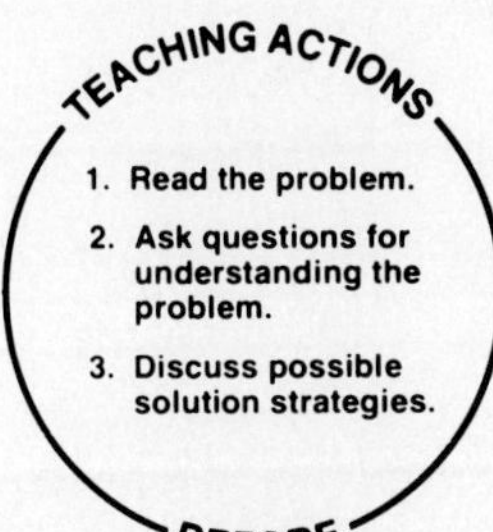

Understanding the Problem

- How many blocks are in his first train? (1)
- How many blocks are in his second train? (7)
- How many blocks are in his third train? (13)

TEACHING ACTIONS

4. Observe students.
5. Give hints as needed for solving the problem.
6. Require students to check back and answer the problem.
7. Give problem extension as needed.

DURING

Planning a Solution

- How many blocks were in the first train? (1) How many more blocks were in the second train? (6)
- How many blocks were in the second train? (7) How many more blocks were in the third train? (6)
- Draw a picture of the fourth train. (19 blocks)

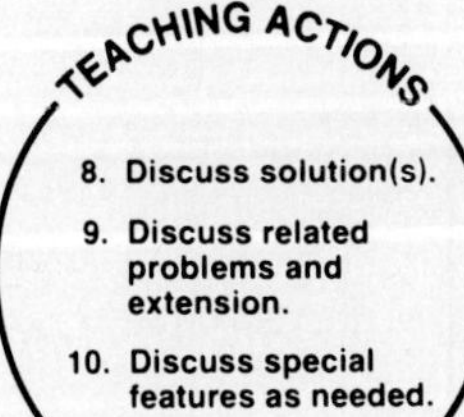

Finding the Answer

Make a Table/Look for a Pattern

Number of Train	Additional Blocks Needed to Build Train	Total Blocks Needed
1	1	1
2	6	1 + 6 = 7
3	6	7 + 6 = 13
4	6	13 + 6 = 19
5	6	19 + 6 = 25
6	6	25 + 6 = 31
7	6	31 + 6 = 37

Ben needs 37 blocks to build his seventh train.

Problem Extension

Ben decides to use a new pattern for building trains. He uses 1 block for his first train, 3 blocks for the second, and 6 blocks for the third. How many blocks will he need for the sixth train? (21)

30 Process Problem

A man was very overweight and his doctor told him to lose 36 kg. If he loses 11 kg the first week, 9 kg the second week, and 7 kg the third week, and he continues losing at this rate, how long will it take him to lose 36 kg? (Hint: Look for a pattern. Then complete the table.)

Week	*Total Kilograms Lost*
1	11
2	11 + 9 = 20
3	20 + 7 = 27
4	
5	

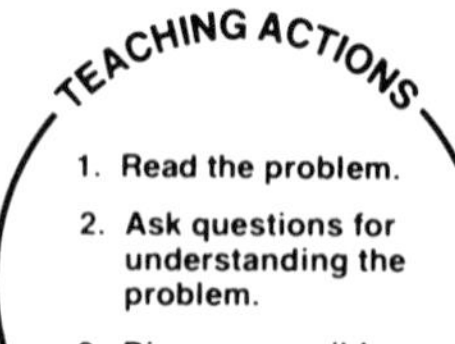

Understanding the Problem

- How much does the man need to lose? (36 kg)
- How much did he lose the first week? (11 kg)
- How much did he lose the second week? (9 kg)
- How much did he lose the third week? (7 kg)

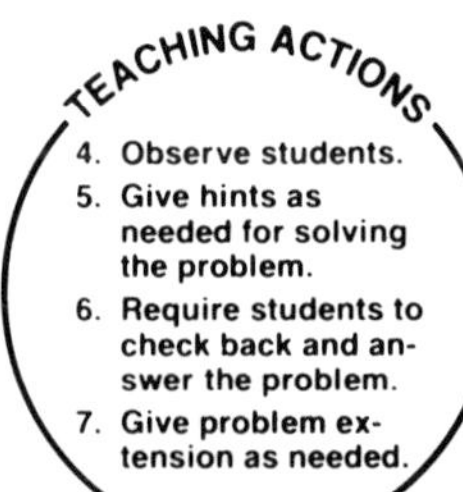

Planning a Solution

- How much less does he lose the second week than the first week? (2 kg)
- How much less does he lose the third week than the second? (2 kg)

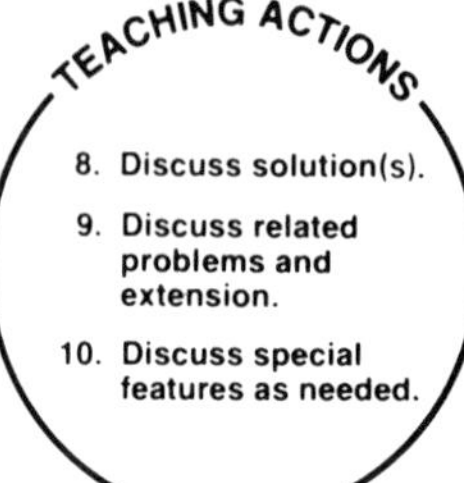

Finding the Answer

Make a Table/Look for a Pattern

Week	*Total Kilograms Lost*
1	11
2	11 + 9 = 20
3	20 + 7 = 27
4	27 + 5 = 32
5	32 + 3 = 35
6	35 + 1 = 36

Pattern: The number of kilograms lost decreases by 2 each week.

It will take the man 6 weeks to lose 36 kg.

Related Problem: 29

Problem Extension

If the man gains his weight back at the rate of 2 kg the first week, 4 kg the second week, 6 kg the third week, and so on, in which week will he have gained back 36 kg? (the sixth)

31 Skill Activity

Write a story problem that this chart would help you to solve.

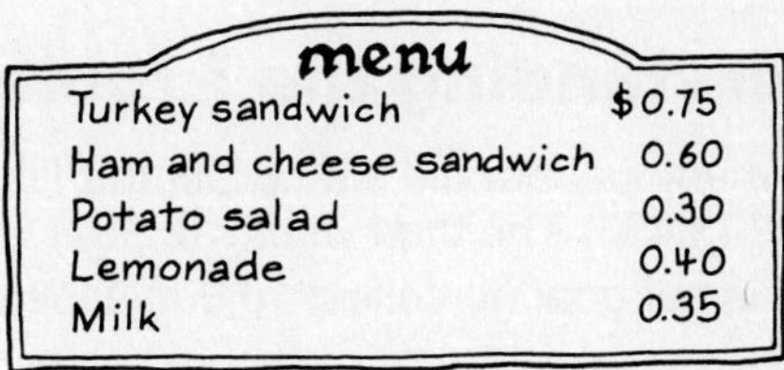

Discussion

Possible problems:

- Irene bought a turkey sandwich, potato salad, and lemonade for her lunch. How much did her lunch cost?
- Tony had $2.00. Did he have enough for 2 ham and cheese sandwiches, potato salad, and milk?

32 One-Step Problem

Joan, Meghan, and Ruth went bowling. How many more points did Meghan score than Joan?

Player	Total
Joan	105
Meghan	181
Ruth	129

Solution

$$\begin{array}{r} 181 \\ -105 \\ \hline 76 \end{array}$$

Meghan scored 76 more points than Joan.

Note: Ruth's score is unnecessary data.

33 Multiple-Step Problem

Our class is collecting newspapers for the newspaper drive. Our goal is to collect 10,000 papers. We collected 768 papers the first week, 3,456 papers the second week, and 2,987 papers the third week. How many more papers do we need to collect?

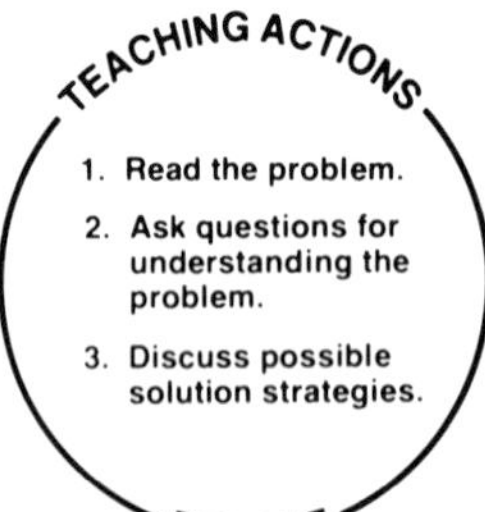

Understanding the Problem

- How many papers did we collect the first week? (768) The second week? (3,456) The third week? (2,987)
- What is our goal (to collect 10,000 papers)

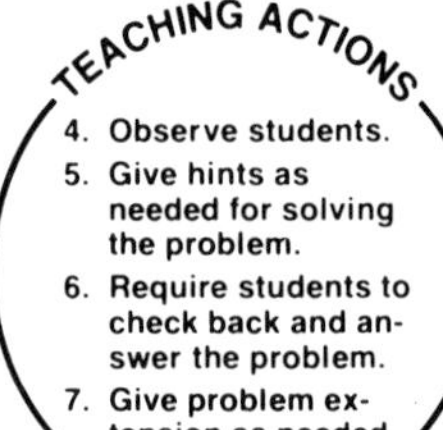

Planning a Solution

- How many have we collected in the first 3 weeks? (768 + 3,456 + 7,211)
- Which operation would you use to find out how many more we need to collect? (subtraction)

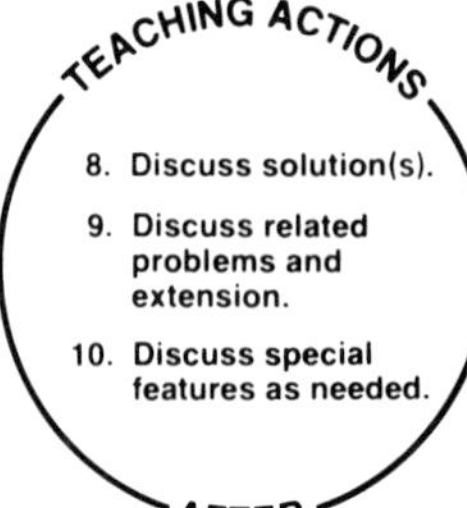

Finding the Answer

Choose the Operations

$$\begin{array}{r} 768 \\ 3{,}456 \\ +2{,}987 \\ \hline 7{,}211 \end{array} \rightarrow \begin{array}{r} 10{,}000 \\ -\ 7{,}211 \\ \hline 2{,}789 \end{array}$$

We need to collect 2,789 more papers.

Related Problems: 23, 18, 13

Problem Extension

If we collect 1,996 papers the next week and 2,006 the following week, how many papers did we collect beyond our goal? (1,213)

34 Process Problem

Leroy, Harold, and Billie left school together. When they left, each of them was wearing the cap that belonged to someone else in that group and the gloves that belonged to yet another person in the group. Harold wore Billie's cap. Whose cap and whose gloves did each boy wear? Complete the chart below to help you use logical reasoning.

	Leroy	Harold	Billie
Cap		Billie's	
Gloves			

TEACHING ACTIONS

1. Read the problem.
2. Ask questions for understanding the problem.
3. Discuss possible solution strategies.

BEFORE

Understanding the Problem

- Who left school together? (Leroy, Harold, and Billie)
- What were they each wearing? (a cap that belonged to someone else and gloves that belonged to yet another person)

TEACHING ACTIONS

4. Observe students.
5. Give hints as needed for solving the problem.
6. Require students to check back and answer the problem.
7. Give problem extension as needed.

DURING

Planning a Solution

- What did Harold wear? (Billie's cap). So, whose gloves did he wear? (Leroy's)
- Whose cap did Leroy wear? (Harold's) Whose gloves? (Billie's)

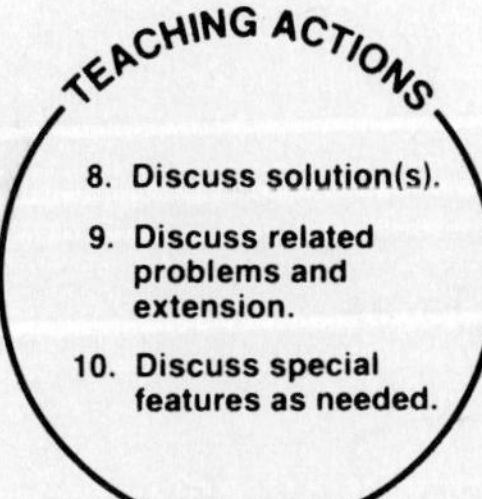

Finding the Answer

Use Logical Reasoning

	Leroy	Harold	Billie
Cap	Harold's	Billie's	Leroy's
Gloves	Billie's	Leroy's	Harold's

Leroy wore Harold's cap and Billie's gloves. Harold wore Billie's cap and Leroy's gloves. Billie wore Leroy's cap and Harold's gloves.

Problem Extension

Suppose Harold wore Leroy's cap. What would the others have to wear? (Harold—Leroy's cap, Billie's gloves; Billie—Harold's cap, Leroy's gloves; Leroy—Billie's cap, Harold's gloves)

35 Process Problem

Four fourth-graders meet on the beach while they are on vacation. Their names are Debbie, Dawn, Cathy, and Becky. They are from California, Illinois, Utah, and Texas. Becky is from California and Dawn is not from Illinois. If Cathy is from Utah, where is Debbie from? (Hint: Complete the chart below to help you use logical reasoning.)

	CA	IL	UT	TX
Debbie				
Dawn		No		
Cathy				
Becky	Yes			

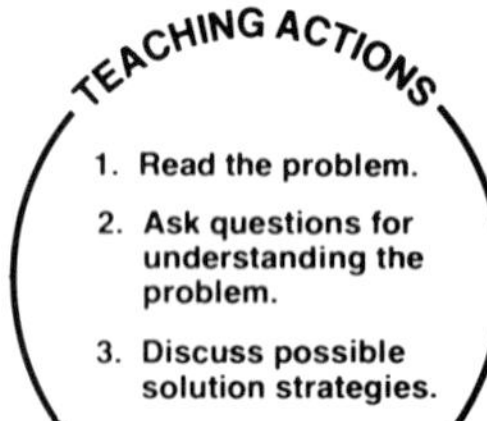

Understanding the Problem

- What are the children's names? (Debbie, Dawn, Cathy, Becky)
- Where are they from? (California, Texas, Illinois, Utah)
- Who is from California? (Becky)
- Who is from Utah? (Cathy)

TEACHING ACTIONS

4. Observe students.
5. Give hints as needed for solving the problem.
6. Require students to check back and answer the problem.
7. Give problem extension as needed.

DURING

Planning a Solution

- Can Debbie or Dawn be from California or Utah? (No)
- Which states and people don't we know about from the problem statement? (Illinois, Texas, Debbie, Dawn)
- If Dawn is not from Illinois, where must she be from? (Texas)

Finding the Answer

Use Logical Reasoning

- We know Becky is from California.
- We know Cathy is from Utah.
- Illinois and Texas remain.
- Dawn is not from Illinois, so she must be from Texas.
- Debbie must be from Illinois.

Becky is from California, Cathy is from Utah, Dawn is from Texas, and Debbie is from Illinois.

Related Problem: 34

Problem Extension

All four girls are sports fans. Their favorite sports are baseball, football, basketball, and tennis. The Californian loves baseball, and the girl from Illinois does not enjoy tennis. If Dawn likes basketball, who is the tennis fan? (Cathy)

36 Skill Activity

This problem contains unnecessary data. Select the data you need to solve the problem. Then find the answer.

Mother's bowling score was 197, Janice's was 213, and Alice's was 181. If Alice and Mother were on a team, what was their total bowling score?

Discussion

You don't need to know Janice's score.

$$\begin{array}{r} 197 \\ +181 \\ \hline 378 \end{array}$$

Mother and Alice scored 378 points together.

37 One-Step Problem

Elliott, Tess, Lisa, and Eric are collecting baseball cards. Elliott has collected 35 cards, Tess has collected 27 cards, Lisa has collected 31 cards, and Eric has collected 29 cards. How many cards have they collected altogether?

Solution

$$\begin{array}{r} 35 \\ 27 \\ 31 \\ +\ 29 \\ \hline 122 \end{array}$$

They have collected 122 baseball cards altogether.

38 Multiple-Step Problem

Justin has collected 32 baseball cards. He began collecting cards when he was 7 and collected the same number of cards every year. He is now 11. How many cards did he collect each year?

TEACHING ACTIONS

1. Read the problem.
2. Ask questions for understanding the problem.
3. Discuss possible solution strategies.

BEFORE

Understanding the Problem

- How many cards does Justin have? (32)
- How old was he when he began collecting? (7)
- How old is he now? (11)
- Did he collect the same number each year? (yes)

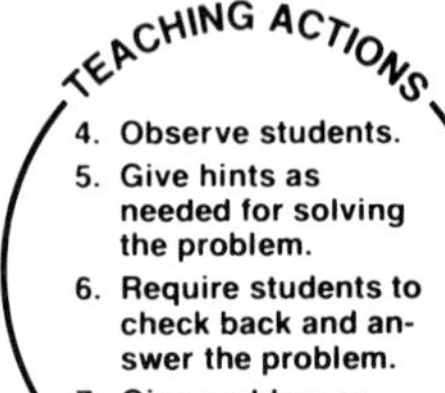

Planning a Solution

- How old was Justin when he began? (7) How old is he now? (11)
- How many years has Justin been collecting baseball cards? (11 − 7 = 4)
- How many cards did he collect altogether? (32)
- Which operation would you use to find the final answer? (division)

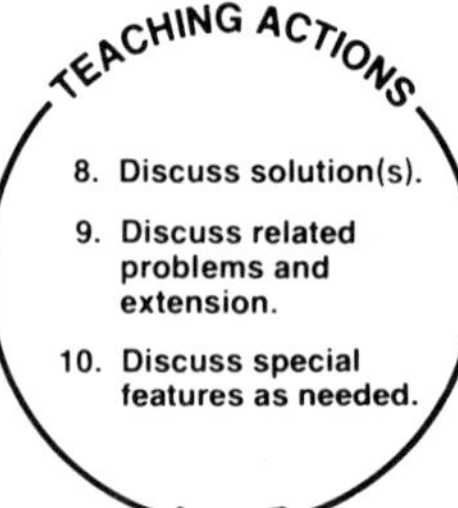

Finding the Answer

Choose the Operations

$$\begin{array}{r} 11 \\ -\ 7 \\ \hline 4 \end{array} \rightarrow 4\overline{)32}\ \text{quotient } 8$$

Justin collected 8 cards each year.

Related Problem: 28

Problem Extension

If Justin continues collecting baseball cards and collects 9 cards each year until he is 16, how many cards will he have by that time? (77)

39 Process Problem

Marty did 2 of these activities. He paid for them with a $10.00 bill. His change was $3.75. What 2 activities did Marty do? (Hint: Make a guess. Then check your guess.)

Activity	Cost
Movies	$3.50
Putt-Putt Golf	3.00
Skating	2.00
Go-Kart Rides	2.75

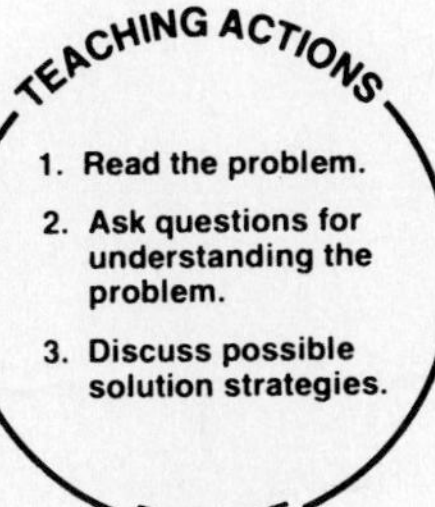

Understanding the Problem

- How many activities did he do? (2)
- How much money did he have? ($10.00)
- What was his change? ($3.75)

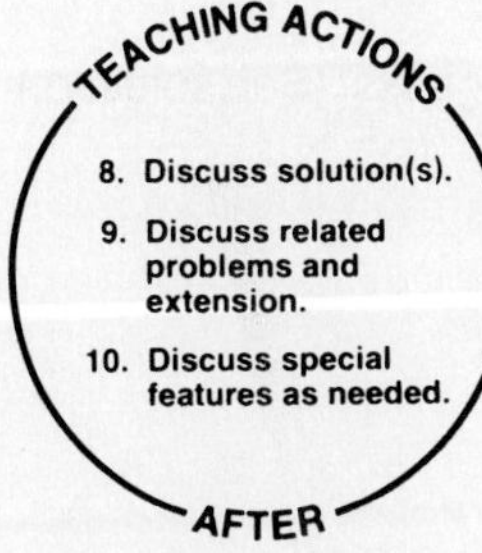

Planning a Solution

- How much money did he have? ($10.00) What was the change? ($3.75)
- How much did he spend? ($6.25)
- If he saw the movies and golfed, how much money would he have spent? ($6.50) Did he do these 2 activities? (No, they cost too much.)

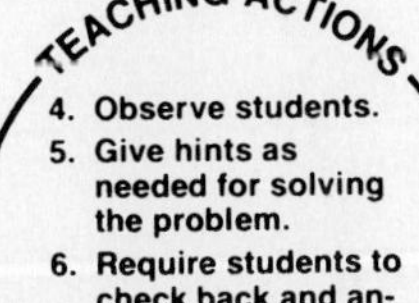

Finding the Answer

Guess and Check

- Try movies and skating—$3.50 + $2.00 = $5.50. (too little)
- Try movies and go-karts—$3.50 + $2.75 = $6.25. (correct)

The activities that Marty did are the movies and the go-kart rides.

Related Problems: 5, 4

Problem Extension

Marty's friend Joe went also, but he was not limited to 2 activities. He took $10.00 and brought back $2.25. What activities did he do? (golf, skating, go-kart rides)

40 Process Problem

Eugene had 33¢ in his pocket. He had 9 coins. He did not have a quarter. What were the coins he had in his pocket?

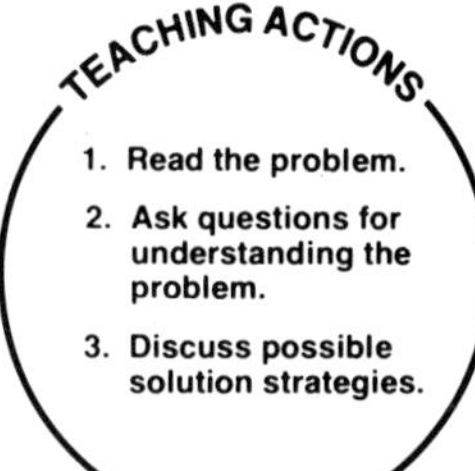

Understanding the Problem

- How much money did Eugene have? (33¢)
- How many coins did he have? (9)
- Does he have a quarter? (no)

TEACHING ACTIONS

4. Observe students.
5. Give hints as needed for solving the problem.
6. Require students to check back and answer the problem.
7. Give problem extension as needed.

DURING

Planning a Solution

- How many pennies must he have? (3)
- If Eugene had 3 dimes, what other change would he have? (3 pennies) How many coins is this? (6) Would this work? (No, we need 9 coins.)
- Try guessing which coins he had. Then check your guess. (See solution.)

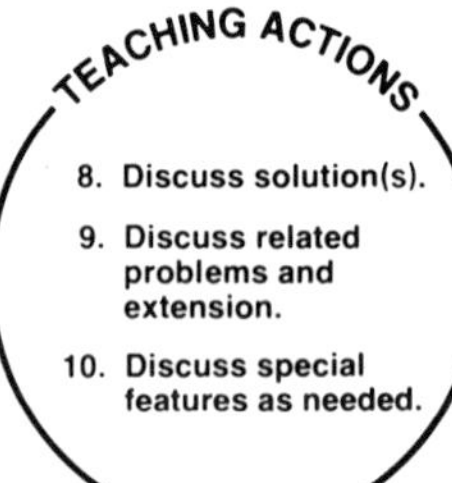

Finding the Answer

Guess and Check

- Try 2D + 2N + 3P = 33¢—7 coins. (too few)
- Try 1D + 4N + 3P = 33¢—8 coins. (too few)
- Try 0D + 6N + 3P = 33¢—9 coins. (correct)

Make a Table

Dimes	Nickels	Pennies	Amount	Number of Coins
3	0	3	33¢	6
2	2	3	33¢	7
1	4	3	33¢	8
0	6	3	33¢	9

Eugene has 6 nickels and 3 pennies in his pocket.

Related Problems: 39, 20, 19, 5, 4

Problem Extension

How can you make 35¢ using 9 coins? (2 dimes, 2 nickels, 5 pennies)

41 Skill Activity

This problem has data missing. Make up appropriate data. Then solve the problem using your data.

Peggy and Camille had a roll of kite string. Peggy took 30 m of the string. How many meters were left?

Discussion

Possible answer:

Peggy and Camille had a roll of kite string that was 75 m long. Peggy took 30 m of the string. How many meters of string were left?

$$\begin{array}{r} 75 \\ -\ \underline{30} \\ 45 \end{array}$$

There were 45 m of string left.

42 One-Step Problem

At Funland there were 4 rides. Each ride costs 20¢ and lasts five minutes. How much does it cost to ride all the rides one time?

Solution

$$\begin{array}{r} 20¢ \\ \times\ \underline{4} \\ 80¢ \end{array}$$

It costs 80¢ to ride all the rides one time.

Note: The fact that each ride lasts five minutes is unnecessary data.

43 Multiple-Step Problem

Manuel paid $1.06 for some wood and $0.72 for some paint to make a new birdhouse. If he paid with a $5.00 bill, what was his change?

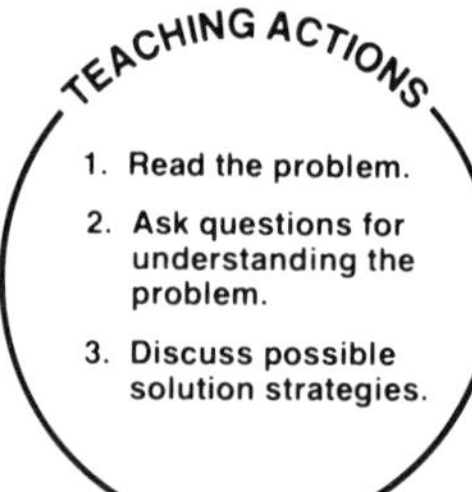

Understanding the Problem

- How much did the wood cost? ($1.06)
- How much did the paint cost? ($0.72)
- How did he pay for it? (with a $5.00 bill)

TEACHING ACTIONS
4. Observe students.
5. Give hints as needed for solving the problem.
6. Require students to check back and answer the problem.
7. Give problem extension as needed.
DURING

Planning a Solution

- What is the total cost of wood and paint? ($1.78)
- Which operation would you use to find his change? (subtraction)

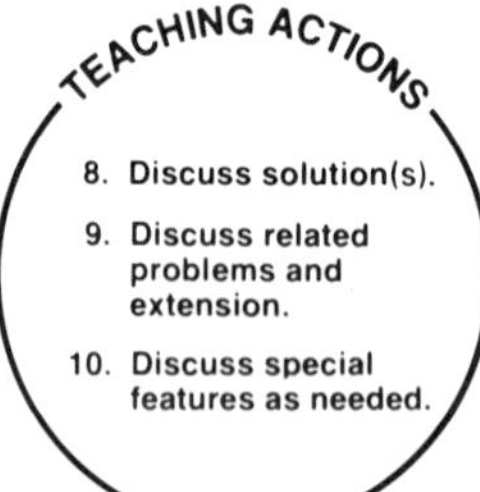

Finding the Answer

Choose the Operations

$$\begin{array}{r} \$1.06 \\ +\ \ .72 \\ \hline \$1.78 \end{array} \quad \rightarrow \quad \begin{array}{r} \$5.00 \\ -\ 1.78 \\ \hline \$3.22 \end{array}$$

Manuel's change was $3.22.

Related Problems: 32, 23, 18, 13

Problem Extension

If Manuel has only $5.00, how much more money does he need to make 3 birdhouses? ($0.34)

44 Process Problem

Pat and Randy had to cut a 10-foot (ft) board into 10 equal pieces. If it takes 1 minute (min) to make each cut, how many minutes will it take them to cut the board into 10 equal pieces? (Hint: Complete the picture.)

TEACHING ACTIONS — BEFORE

1. Read the problem.
2. Ask questions for understanding the problem.
3. Discuss possible solution strategies.

Understanding the Problem

- How long does it take to make each cut? (1 min)
- How long is the board? (10 ft)
- How many pieces do they want to cut it into? (10)

TEACHING ACTIONS — DURING

4. Observe students.
5. Give hints as needed for solving the problem.
6. Require students to check back and answer the problem.
7. Give problem extension as needed.

Planning a Solution

- If Pat and Randy make 1 cut, how many pieces do they have? (2)
- How long did it take? (1 min)
- How many pieces do they have when they make 2 cuts? (3) How many minutes did it take? (2)
- Try 4 cuts. (5 pieces, 4 min)

TEACHING ACTIONS — AFTER

8. Discuss solution(s).
9. Discuss related problems and extension.
10. Discuss special features as needed.

Finding the Answer

Draw a Picture

1 2 3 4 5 6 7 8 9

It takes 9 min to cut a 10-ft board into 10 equal pieces.

Related Problems: 10, 9

Problem Extension

How long does it take to cut a board into 100 pieces? (99 min) Into *n* pieces? ($n - 1$ min)

45 Process Problem

A building has 6 stories, each the same height. It takes 10 seconds (s) for the elevator to get from the first floor to the third floor. How many seconds does it take for the elevator to get from the first floor to the sixth floor?

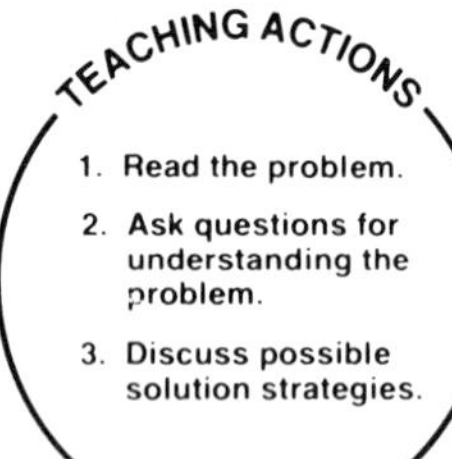

Understanding the Problem

- How many stories are there in the building? (6)
- How long does it take to get from the first to the third floor? (10 s)
- How many stories is that? (2)

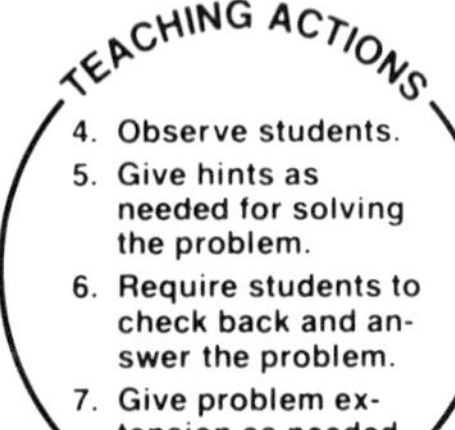

Planning a Solution

- How many seconds would it take to get from the first to the second floor if it takes 10 s to get from the first to the third? (10 ÷ 2 = 5)
- How long would it take to get to the fourth floor from the third floor? (5 s)
- Try drawing a picture. (See solution.)

Finding the Answer

Draw a Picture

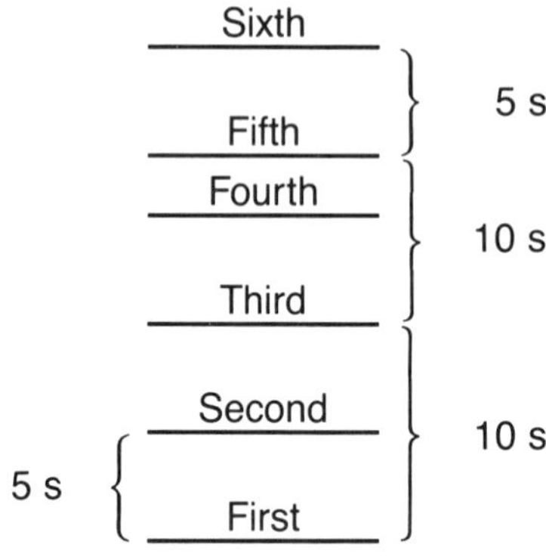

It takes 25 s to get from the first floor to the sixth floor.

Related Problems: 44, 10, 9

Problem Extension

Move 1 block to another stack so that the sums of all the stacks are equal. (Move the 9 to the first stack; now all stacks equal 18.)

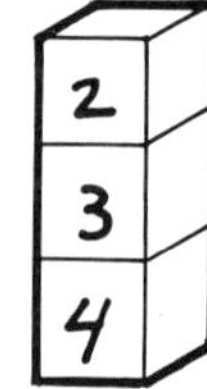

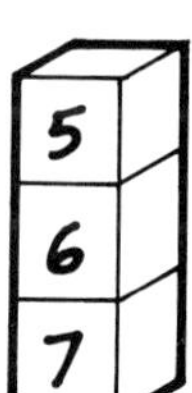

46 Skill Activity

Identify the operation or operations needed to solve each problem. Then choose one problem and solve it.

1. Roger buys baseball cards for 10¢ each and sells them for 17¢ each. How much does he make on each card he sells?

2. Chang hit 23 home runs last year and 13 home runs the year before. If he makes 27 home runs this year, how many home runs will he have made in 3 years?

Discussion

Possible answers:

1. Use subtraction to solve this problem.

$$\begin{array}{r} 17¢ \\ -10¢ \\ \hline 7¢ \end{array}$$

Roger will make 7¢ on each card he sells.

2. Use addition to solve this problem.

$$\begin{array}{r} 23 \\ 13 \\ +27 \\ \hline 63 \end{array}$$

Chang will have made 63 home runs in 3 years.

47 One-Step Problem

Rosa had \$5.00. She spent \$3.50 for a basketball ticket and some popcorn. How much did she have left?

Discussion

$$\begin{array}{r} \$5.00 \\ -\ 3.50 \\ \hline \$1.50 \end{array}$$

Rosa had \$1.50 left.

48 Multiple-Step Problem

Betty's allowance is \$3.60. She spends \$1.20 of it on a record and \$1.44 of it on a book. How much of her allowance does she have left?

TEACHING ACTIONS

1. Read the problem.
2. Ask questions for understanding the problem.
3. Discuss possible solution strategies.

BEFORE

Understanding the Problem

- What is Betty's allowance? (\$3.60)
- How much did she pay for the record? (\$1.20)
- How much did she pay for the book? (\$1.44)

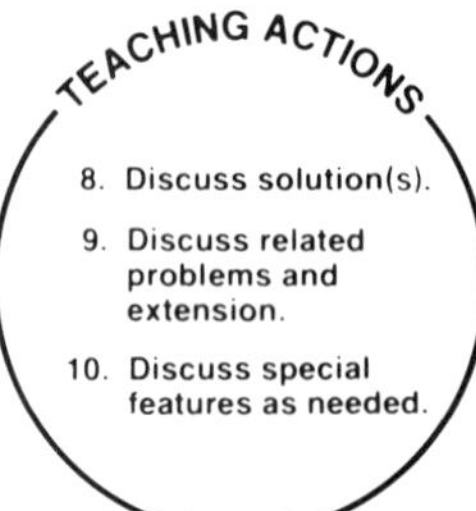

Planning a Solution

- What was the cost of the record and book? (\$1.20 + \$1.44 = \$2.64)
- Which operation would you use to find out how much of her allowance she has left? (subtraction)

TEACHING ACTIONS

8. Discuss solution(s).
9. Discuss related problems and extension.
10. Discuss special features as needed.

AFTER

Finding the Answer

Choose the Operation

$$\begin{array}{r} \$1.20 \\ +\ .44 \\ \hline \$2.64 \end{array} \rightarrow \begin{array}{r} \$3.60 \\ -\ 2.64 \\ \hline \$\ 0.96 \end{array}$$

Betty has \$0.96 of her allowance left.

Related Problems: 43, 33, 23, 18, 13

Problem Extension

How much more allowance would Betty need to get to buy 2 records and 2 books? (\$1.68)

49 Process Problem

Andrew wants to participate in 2 running events in the track meet. He can select from the 50-m dash, 75-m dash, 100-m dash, and 400-m relay. How many different choices does Andrew have? (Hint: Make an organized list.)

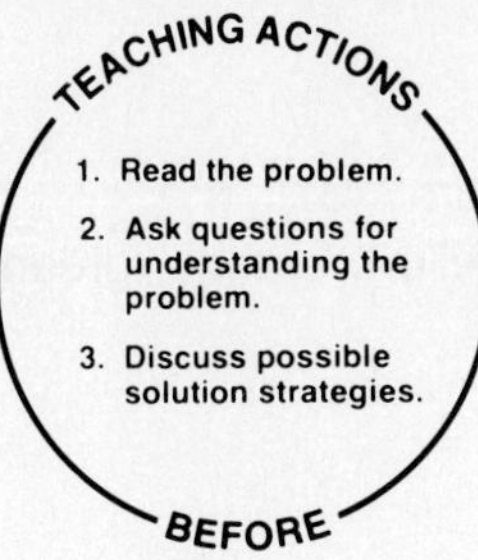

Understanding the Problem

- How many events does Andrew want to enter? (2)
- What events can he select from? (50-m, 75-m, 100-m, 400-m)

TEACHING ACTIONS

4. Observe students.
5. Give hints as needed for solving the problem.
6. Require students to check back and answer the problem.
7. Give problem extension as needed.

DURING

Planning a Solution

- Can Andrew choose the 50-m and the 75-m races? (yes) 50-m and 100-m races? (yes)
- Make an organized list to find all the choices.

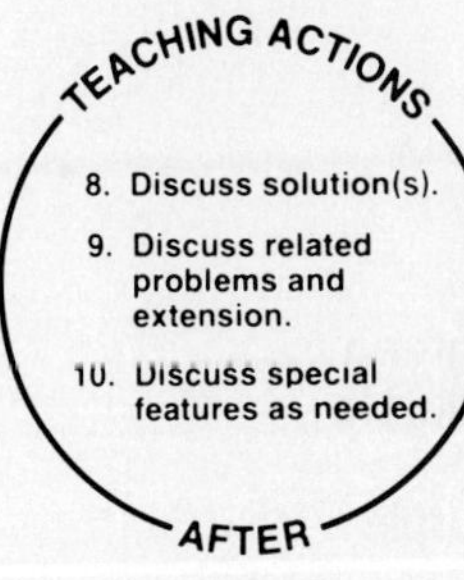

Finding the Answer

Make an Organized List

50— 75
50—100
50—400

75—100
75—400

100—400

Andrew has 6 choices.

Related Problems: 15, 14

Problem Extension

Andrew also wants to enter 2 field events in addition to the 2 running events. He can choose from the high jump, standing broad jump, shotput, and discus. How many different choices does Andrew now have?

50-m, 75-m, hj, sbj
50-m, 75-m, hj, shotput
50-m, 75-m, hj, discus
50-m, 75-m, sbj, shotput
50-m, 75-m, sbj, discus
50-m, 75-m, shotput, discus
and so on

Andrew now has 36 different choices.

50 Process Problem

Ms. Davis asked her students each to draw a picture describing their hobby. She asked them to use only 3 colors. Patrick had a box of crayons containing the colors red, blue, yellow, orange, green, and purple. How many different ways can Patrick use his crayons to draw his picture? (Hint: Make an organized list.)

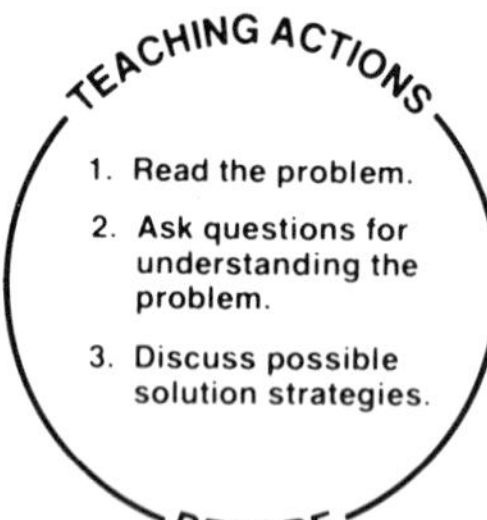

Understanding the Problem

- How many colors could Patrick use on his picture? (3)
- What colors does Patrick have? (red, blue, yellow, orange, green, purple)
- How many colors is that? (6)

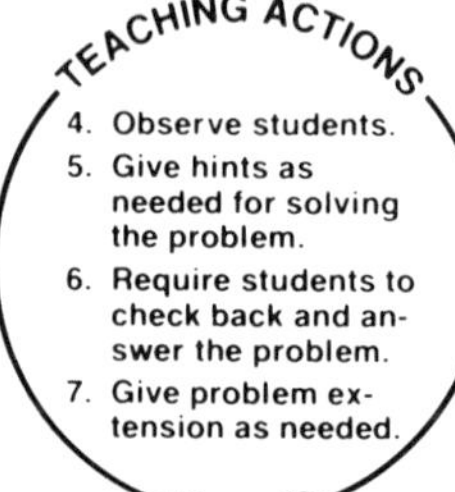

Planning a Solution

- If Patrick chooses red and blue, how many choices does he have for the third crayon? (4) Name one of the choices. (yellow, orange, green, or purple)
- If one of the combinations Patrick can choose is red, blue, and yellow, is another one yellow, red, and blue? (no)

Finding the Answer

Make an Organized List

Red, blue, yellow
Red, blue, orange
Red, blue, green
Red, blue, purple

Red, yellow, orange
Red, yellow, green
Red, yellow, purple

Red, orange, green
Red, orange, purple
Red, green, purple

Blue, yellow, orange
Blue, yellow, green
Blue, yellow, purple

Blue, orange, green
Blue, orange, purple

Blue, green, purple

Yellow, orange, green
Yellow, orange, purple
Yellow, green, purple

Orange, green, purple

Patrick can use his crayons 20 different ways.

Related Problems: 49, 15, 14

Problem Extension

If Ms. Davis allowed Patrick to use only 2 colors, how many choices would he have? (15)

51 Skill Activity

Determine whether the answers given for these problems are reasonable. If an answer is not reasonable, explain why.

1. Amber went for a hike on each Saturday in May. She hiked 4 miles on the first Saturday, 7 miles on the second Saturday, and 9 miles on each of the next two Saturdays in May. How far did Amber hike altogether in May?
   Answer: Amber hiked 20 miles in May.

2. Charles had 36 seashells that he collected at the beach. He gave each of his four friends an equal number of seashells. How many did each boy receive?
   Answer: Each boy received 9 seashells.

Discussion

Possible answers:

1. This answer is not reasonable because Amber hiked 9 miles on *each* of the last 2 Saturdays in May.

2. This answer is reasonable because $36 \div 4 = 9$.

52 One-Step Problem

There are 5 rows of desks in Mr. Sanchez's fourth grade room and each row has 7 desks in it. How many desks are there in Mr. Sanchez's room?

Solution

$$\begin{array}{r} 5 \\ \times\ 7 \\ \hline 35 \end{array}$$

There are 35 desks in Mr. Sanchez's room.

53 Multiple-Step Problem

I sold 9 blue marbles and 8 white marbles. I sold the blue ones for $0.07 each and the white ones for $0.05 each. How much money did I get for all the marbles?

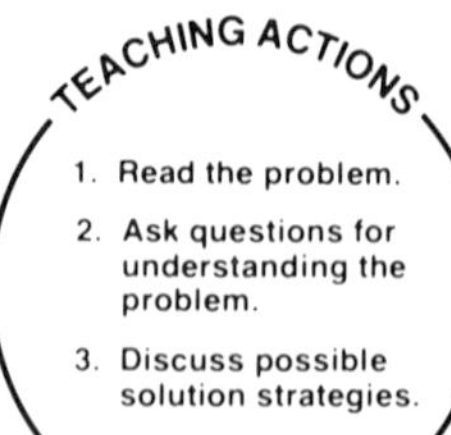

Understanding the Problem

- How many blue marbles did I sell? (9) For how much did I sell each blue marble? ($0.07)
- How many white marbles did I sell? (8) For how much did I sell each white marble? ($0.05)

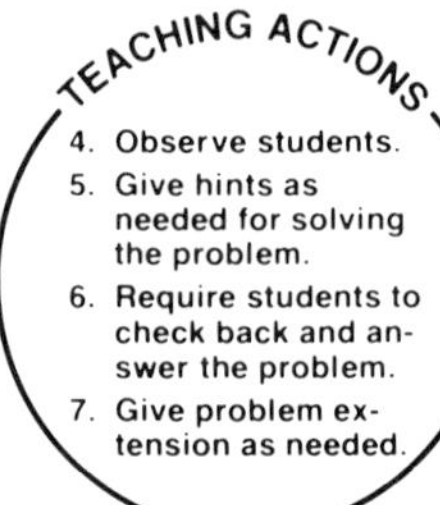

Planning a Solution

- If 1 blue marble costs $0.07, how much do 9 blue marbles cost? ($0.07 × 9 = $0.63)
- What is the cost of 8 white marbles? ($0.05 × 8 = $0.40)

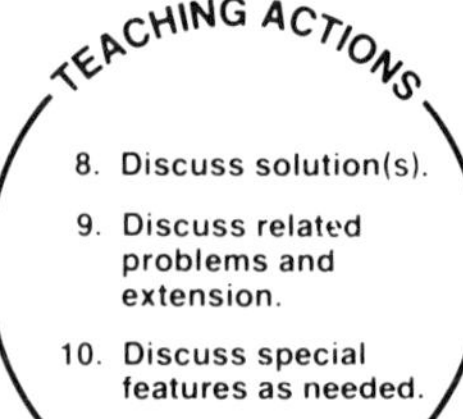

Finding the Answer

Choose the Operations

$$\begin{array}{r} \$0.07 \\ \times\ 9 \\ \hline \$0.63 \end{array} \rightarrow \begin{array}{r} \$0.05 \\ \times\ 8 \\ \hline \$0.40 \end{array} \rightarrow \begin{array}{r} \$0.63 \\ +0.40 \\ \hline \$1.03 \end{array}$$

I got $1.03 for the marbles.

Related Problem: 8

Problem Extension

How much would I make if I sold the blue marbles for $0.10 each and the white ones for $0.09 each? ($1.62)

61 Skill Activity

Write a story problem that this table would help you solve.

Home Phone Rates	
First Minute	**Each Additional Minute**
23¢	**12¢**

Discussion

Possible problem:

How much would it cost to make a 3-min call? (47¢)

62 One-Step Problem

Dean and Jan went bowling. If their team score was 149, what was Dean's score?

Total	Player
?	Dean
97	Jan

Solution

$$\begin{array}{r} 149 \\ -\ 97 \\ \hline 52 \end{array}$$

Dean's score was 52.

63 Multiple-Step Problem

How much money do I need to buy 3 tennis balls at $0.50 each and 1 racket that costs $6.00?

TEACHING ACTIONS

1. Read the problem.
2. Ask questions for understanding the problem.
3. Discuss possible solution strategies.

BEFORE

Understanding the Problem

- How much do tennis balls cost each? ($0.50)
- How many tennis balls do I want to buy? (3)
- How much does it cost for 1 racket? ($6.00)

TEACHING ACTIONS

4. Observe students.
5. Give hints as needed for solving the problem.
6. Require students to check back and answer the problem.
7. Give problem extension as needed.

DURING

Planning a Solution

- If one ball costs $0.50, what operation would you use to find the cost of 3 balls? (multiplication)
- How do you find the total cost of balls and racket? (add)

TEACHING ACTIONS

8. Discuss solution(s).
9. Discuss related problems and extension.
10. Discuss special features as needed.

AFTER

Finding the Answer

Choose the Operations

$$\begin{array}{r} \$0.50 \\ \times\ 3 \\ \hline \$1.50 \end{array} \rightarrow \begin{array}{r} \$6.00 \\ +\ 1.50 \\ \hline \$7.50 \end{array}$$

I need $7.50.

Related Problems: 38, 28

Problem Extension

If I have $10.00, how many more tennis balls can I buy? (5 more)

64 Process Problem

José used 6 blocks to build this staircase with 3 steps. How many blocks will José need to make a 6-step staircase? (Hint: Make a table and look for a pattern.)

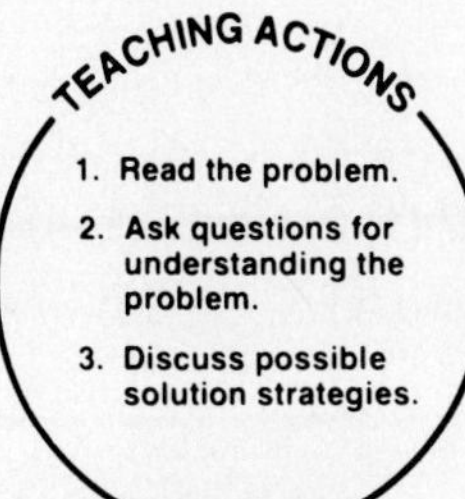

Understanding the Problem

- How many blocks are used to build a 3-step staircase? (6)
- Do you know how many blocks are used to make a 6-step staircase? (No, that is what we are trying to find out.)

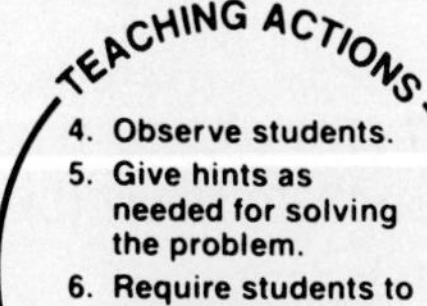

Planning a Solution

- How many blocks were used to build the first step? (1)
- How many new blocks were used for the second step? (2)
- How many new blocks would be needed for the fourth step? (4) What would be the total number of blocks used to build a staircase with 4 steps? (10)

Finding the Answer

Make a Table/Look for a Pattern

Steps in Staircase	***Blocks Needed to Build New Steps***	***Total Blocks Needed***
1	**1**	**1**
2	**2**	**1 + 2 = 3**
3	**3**	**1 + 2 + 3 = 6**
4	**4**	**1 + 2 + 3 + 4 = 10**
5	**5**	**1 + 2 + 3 + 4 + 5 = 15**
6	**6**	**1 + 2 + 3 + 4 + 5 + 6 = 21**

Pattern: The number of new blocks needed increases by 1 with each new step. The total number of blocks needed for *n*th step is the sum of the number 1 through *n*.

It would take 21 blocks to build a 6-step staircase.

Related Problems: 55, 54, 40, 30, 29

Problem Extension

How many steps would there be in a staircase using 78 blocks? (12)

65 Process Problem

Earl played a game using the figure below. First he covered the section numbered 1. Then he covered the sections numbered 1 and 2. Next he covered the sections numbered 1 and 4. What sections would he cover on his seventh round?

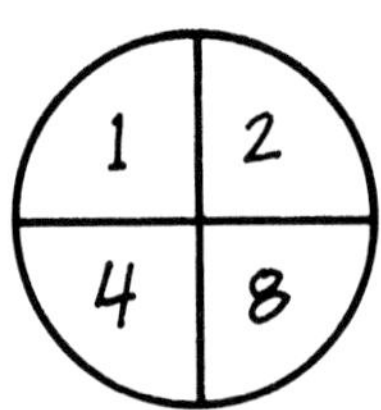

TEACHING ACTIONS

1. Read the problem.
2. Ask questions for understanding the problem.
3. Discuss possible solution strategies.

BEFORE

Understanding the Problem

- What numbers are in the circle? (1, 2, 4, 8)
- What number(s) did he cover first? (1) Second? (1, 2) Next? (1, 4)

TEACHING ACTIONS

4. Observe students.
5. Give hints as needed for solving the problem.
6. Require students to check back and answer the problem.
7. Give problem extension as needed.

DURING

Planning a Solution

- What is the sum of the numbers he covered first? (1)
- What is the sum of the numbers he covered second? (3) Next? (5)
- Make a table and look for a pattern. (See solution.)

TEACHING ACTIONS

8. Discuss solution(s).
9. Discuss related problems and extension.
10. Discuss special features as needed.

AFTER

Finding the Answer

Make a Table/Look for a Pattern

Round	Sum
First	1
Second	1 + 2 = 3
Third	1 + 4 = 5
Fourth	1 + 2 + 4 = 7
Fifth	1 + 8 = 9
Sixth	1 + 2 + 8 = 11
Seventh	1 + 4 + 8 = 13

Pattern: The sum of the numbers increases by 2 in each round.

Earl would cover the 1, 4, and 8 on his seventh round.

Related Problems: 64, 55, 54, 40, 30

Problem Extension

If he covered the 2 first, then the 4, then the 2 and the 4, what numbers would he cover on his seventh round? (2, 4, 8)

66 Skill Activity

These problems have the numbers omitted. Identify the operation or operations needed to solve each problem.

1. Monique gets an allowance each week. She saves part of it. How much does Monique have left over to spend?

2. Katie hit some home runs, Dana hit some home runs, and Lizzie hit some home runs. How many home runs did they hit altogether?

Discussion

Possible answers:

1. Use subtraction to solve this problem.

2. Use addition to solve this problem.

67 One-Step Problem

When Amy walked 9 blocks, she was halfway to Kari's house. How many blocks is it from Amy's house to Kari's house?

Solution

$$\begin{array}{r} 9 \\ \times\ 2 \\ \hline 18 \end{array}$$

It is 18 blocks from Amy's house to Kari's house.

68 Multiple-Step Problem

Mr. and Mrs. Kim and their 2 children went to the game. How much did they pay in all?

Adults	$2
Children	$1

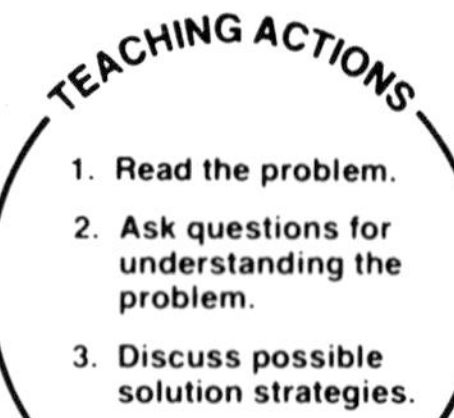

Understanding the Problem

- How much does 1 adult ticket cost? ($2)
- How many adults went to the game? (2)
- How much does 1 children's ticket cost? ($1)
- How many children went to the game? (2)

TEACHING ACTIONS
4. Observe students.
5. Give hints as needed for solving the problem.
6. Require students to check back and answer the problem.
7. Give problem extension as needed.
DURING

Planning a Solution

- How much did they pay for 2 adults? ($4)
- How much did they pay for 2 children? ($2)
- Which operation would you use to find the total cost? (addition)

TEACHING ACTIONS
8. Discuss solution(s).
9. Discuss related problems and extension.
10. Discuss special features as needed.
AFTER

Finding the Answer

Choose the Operations

$$\begin{array}{r} \$2 \\ \times\ 2 \\ \hline \$4 \end{array} \rightarrow \begin{array}{r} \$1 \\ \times\ 2 \\ \hline \$2 \end{array} \rightarrow \begin{array}{r} \$4 \\ +\ 2 \\ \hline \$6 \end{array}$$

Mr. and Mrs. Kim and their 2 children paid $6 in all.

Related Problems: 53, 8

Problem Extension

What was their change if they paid for their tickets with a $10 bill? ($4)

69 Process Problem

Laura, Anita, Vera, and Lois played in a backgammon tournament. Laura lost to Lois in the first-round game. Vera won 1 game and lost 1 game. Vera played Lois in the second round. Who won the tournament? (Hint: Make a chart to help you use logical reasoning.)

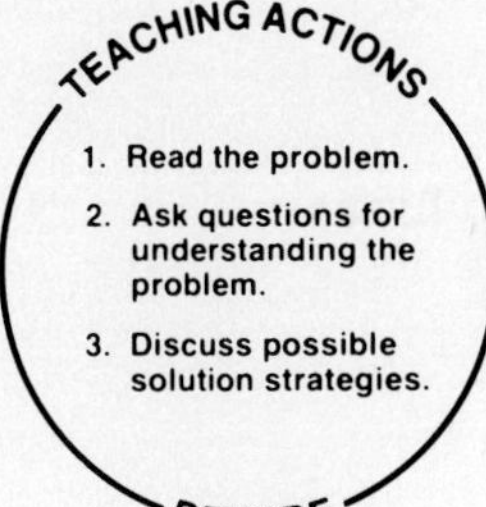

Understanding the Problem

- Who played in the backgammon tournament? (Laura, Anita, Vera, and Lois)
- Who lost in the first round? (Laura) Whom did she play? (Lois)
- How many games did Vera win? (1)

TEACHING ACTIONS

4. Observe students.
5. Give hints as needed for solving the problem.
6. Require students to check back and answer the problem.
7. Give problem extension as needed.

DURING

Planning a Solution

- Whom did Vera play in the first round? (Anita)
- Did Vera win the first round? (Yes)
- Whom did Vera play in the second round? (Lois)

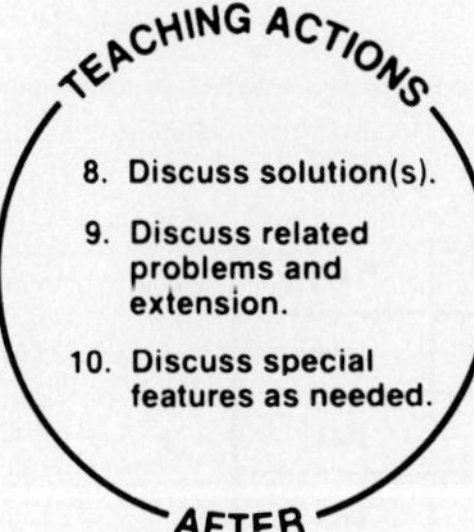

Finding the Answer

Use Logical Reasoning

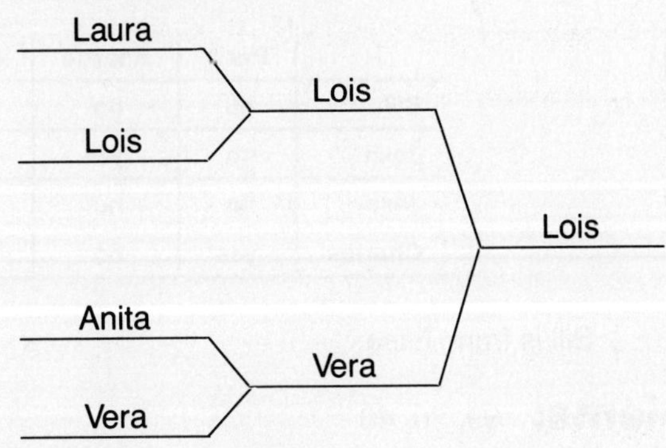

Lois won the tournament.

Related Problems: 35, 34

Problem Extension

Suppose 8 people played in a backgammon tournament. If a person was eliminated when he or she lost a game, how many games would have to be played to determine a champion? (7)

70 Process Problem

Bill, Josh, Jason, and Charles are boys from different cities who met at a party. They are from the cities of Paris, Seattle, Knoxville, and Tulsa. Charles is from Paris, and Josh is not from Tulsa. If Jason is from Knoxville, where is Bill from?

TEACHING ACTIONS

1. Read the problem.
2. Ask questions for understanding the problem.
3. Discuss possible solution strategies.

BEFORE

Understanding the Problem

- Who are the boys? (Bill, Josh, Jason, Charles)
- Where are they from? (Paris, Seattle, Knoxville, Tulsa)

TEACHING ACTIONS

4. Observe students.
5. Give hints as needed for solving the problem.
6. Require students to check back and answer the problem.
7. Give problem extension as needed.

DURING

Planning a Solution

- Where is Charles from? (Paris)
- Where is Jason from? (Knoxville)
- Josh is not from where? (Tulsa)
- Make a chart to help you use logical reasoning. (See solution.)

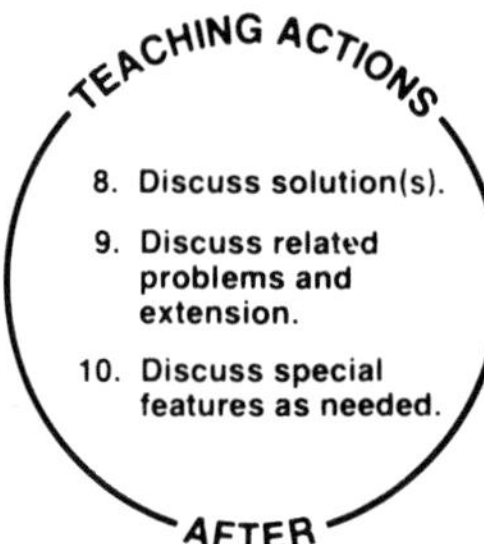

Finding the Answer

Use Logical Reasoning

	Paris	Seattle	Knoxville	Tulsa
Bill	no	no	no	yes
Josh	no	yes	no	no
Jason	no	no	yes	no
Charles	yes	no	no	no

Bill is from Tulsa.

Related Problems: 69, 35, 40

Problem Extension

Arrange these 4 colored rings in order from smallest to largest using the following facts: The yellow is smaller than the red. The blue is larger than the white. The white is smaller than the yellow. The red is 3 rings larger than the white. The white is 2 rings away from the blue. (white, yellow, blue, red)

71 Skill Activity

Write a story problem that you can solve by using the following number sentence:

59 − 43 = ?

Discussion

Possible answer:

Andrew collected 59 bottle caps and Sally collected 43. How many more bottle caps did Andrew collect than Sally?

72 One-Step Problem

Anaheim Stadium has a seating capacity of 43,204. Yankee Stadium has a seating capacity of 53,160. How many more people can be seated in Yankee Stadium than in Anaheim Stadium?

Solution

$$\begin{array}{r} 53{,}160 \\ -\,43{,}204 \\ \hline 9{,}956 \end{array}$$

Yankee Stadium will seat 9,956 more people than Anaheim Stadium.

73 Multiple-Step Problem

Jill bought 3 records and gave the clerk a $10.00 bill. How much change should she receive?

TEACHING ACTIONS

1. Read the problem.
2. Ask questions for understanding the problem.
3. Discuss possible solution strategies.

BEFORE

Understanding the Problem

- How much did each record cost? ($1.75)
- How many records did she buy? (3)
- What did Jill give the clerk? (a $10.00 bill)

TEACHING ACTIONS

4. Observe students.
5. Give hints as needed for solving the problem.
6. Require students to check back and answer the problem.
7. Give problem extension as needed.

DURING

Planning a Solution

- If one record costs $1.75, how much do 3 records cost? ($1.75 × 3 = $5.25)
- Which operation would you use to find out how much change Jill should receive? (subtraction)

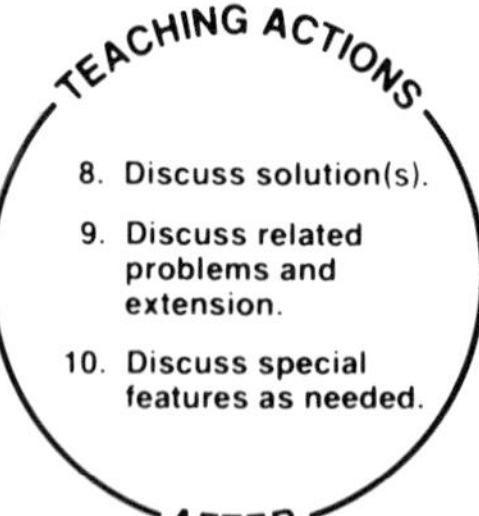

Finding the Answer

Choose the Operations

$$\begin{array}{r} \$1.75 \\ \times \quad 3 \\ \hline \$5.25 \end{array} \rightarrow \begin{array}{r} \$10.00 \\ - \quad 5.25 \\ \hline \$\ 4.75 \end{array}$$

Jill should receive $4.75 in change.

Related Problems: 63, 38, 28

Problem Extension

What is the maximum number of records Jill could buy with her $10.00 bill? (5) What would her change be then? ($1.25)

74 Process Problem

The sum of the ages of three children is 22. Susan is the oldest. Mark is not the youngest. Bobby is 6 years younger than the oldest, who is 10 years old. What is Mark's age?

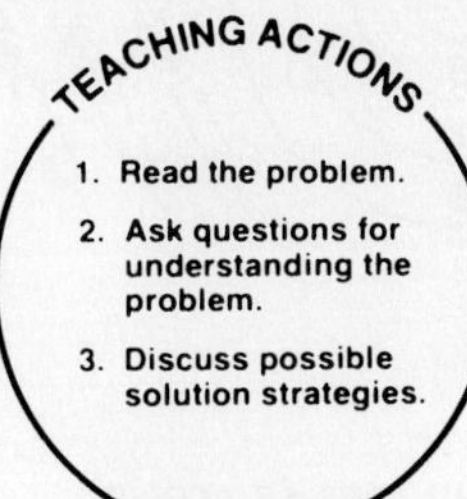

Understanding the Problem

- Who are the three children? (Susan, Mark, Bobby)
- What do we know about Susan? (oldest) How old? (10)
- What do we know about Mark? (not youngest)
- What do we know about Bobby? (6 years younger than the oldest, who is 10)
- What is the sum of their ages? (22)

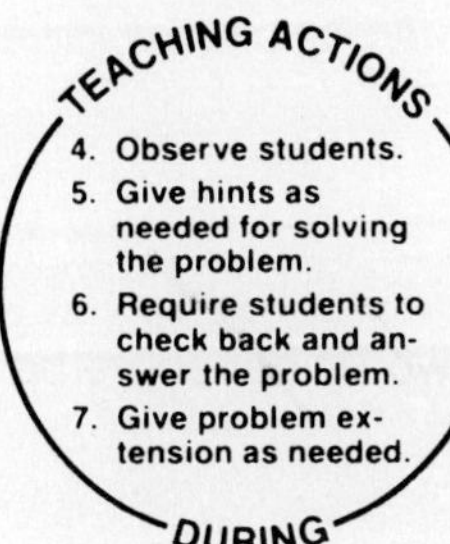

Planning a Solution

- How old is Susan? (10)
- How old is Bobby? ($10 - 6 = 4$)
- Write what you know. Then use logical reasoning to find the other ages. (See solution.)

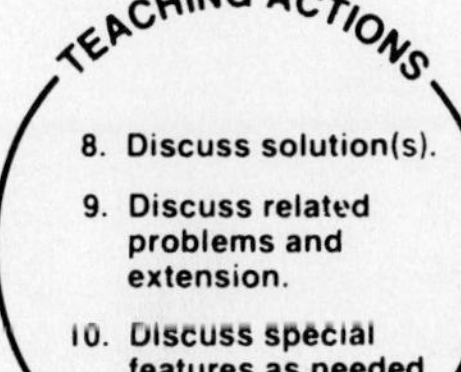

Finding the Answer

Use Logical Reasoning/Guess and Check

Susan—10

Mark—? $10 + ? + 4 = 22$

Bobby—4

Mark is 8 years old.

Related Problems: 70, 69, 40, 39, 35

Problem Extension

I am twice as old as my brother. The sum of our ages is 15. What are our ages? (I am 10 and my brother is 5.)

75 Process Problem

Jane scored exactly 55 points using 3 darts. Show how she might have scored.

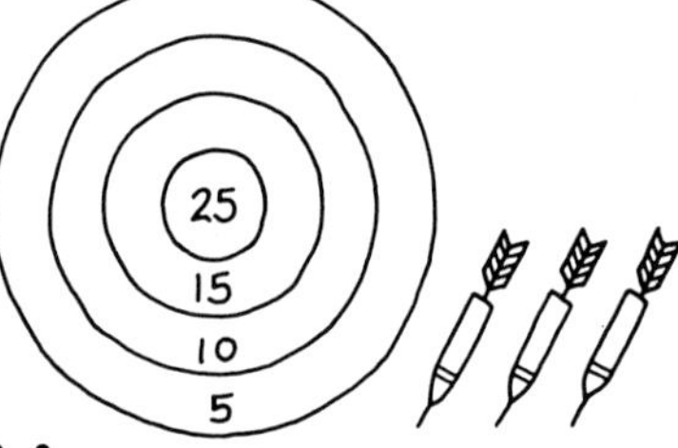

TEACHING ACTIONS

1. Read the problem.
2. Ask questions for understanding the problem.
3. Discuss possible solution strategies.

BEFORE

Understanding the Problem

- How many points did Jane score? (55)
- How many darts did she use? (3)
- What are the scores showing on the dart board? (25, 15, 10, 5)

TEACHING ACTIONS

4. Observe students.
5. Give hints as needed for solving the problem.
6. Require students to check back and answer the problem.
7. Give problem extension as needed.

DURING

Planning a Solution

- If Jane threw 3 darts in the 5-point circle, what would her score be? (15)
- Is this the total points we need? (no)
- Make an organized list to show how she might have scored 55 points. (See solution.)

TEACHING ACTIONS

8. Discuss solution(s).
9. Discuss related problems and extension.
10. Discuss special features as needed.

AFTER

Finding the Answer

Guess and Check/Make an Organized List

5	3	0	0	0	2	0	0	1	0	1	0
10	0	3	0	0	1	2	0	0	1	1	1
15	0	0	3	0	0	1	2	0	2	1	1
25	0	0	0	3	0	0	1	2	0	0	0
SCORE	15	30	45	75	20	35	55	55	40	30	50
	(no)	(no)	(no)	(no)	(no)	(no)	(yes)	(yes)	(no)	(no)	(no)

To get a total of 55, Jane might have scored two 15s, and one 25 or two 25s and one 5.

Related Problems: 74, 50, 49, 40, 39

Problem Extension

How can Jane score 100 points using 8 darts? (four 5s, two 15s, and two 25s; five 5s and three 25s; five 15s, two 10s, one 5; one 25, four 15s, three 5s; or two 5s, four 10s, and two 25s)

76 Skill Activity

This problem has data missing. Make up appropriate data. Then find the answer using your data.

Erica and Sean collected gum wrappers. Erica collected 425 wrappers and Sean collected the rest. How many gum wrappers did Erica and Sean collect altogether?

Discussion

Possible answer:

Sean collected 536 gum wrappers.

$$\begin{array}{r} 425 \\ +\ 536 \\ \hline 961 \end{array}$$

They collected 961 wrappers altogether.

77 One-Step Problem

Sasha and her brother each found 42 shells on their vacation at the beach. How many shells did Sasha and her brother find?

Solution

$$\begin{array}{r} 42 \\ \times\ 2 \\ \hline 84 \end{array}$$

Sasha and her brother found 84 shells.

78 Multiple-Step Problem

It is 4,140 km from New York to San Francisco. The airplane flew 825 km the first hour and 865 km the second hour. How much farther does the plane have to fly before it reaches San Francisco?

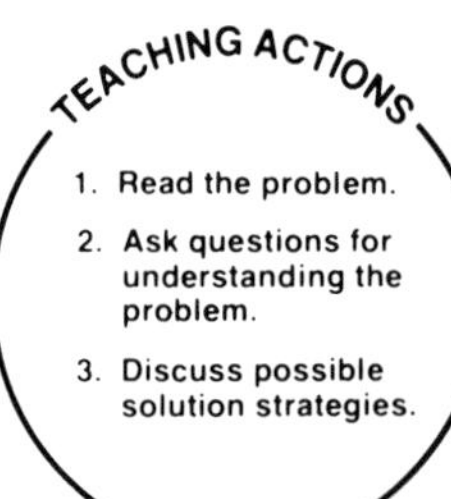

Understanding the Problem

- How far is it from New York to San Francisco? (4,140 km)
- How far did the plane fly the first hour? (825 km)
- How far did the plane fly the second hour? (865 km)

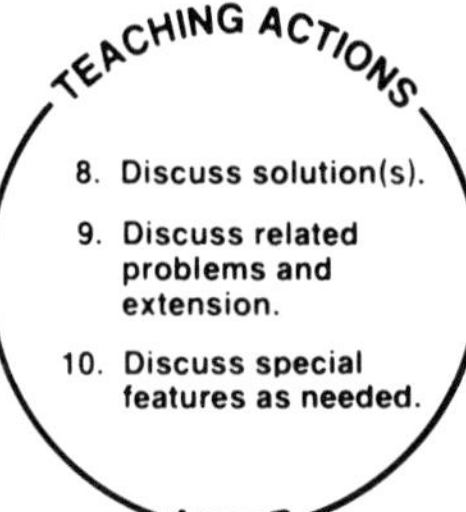

Planning a Solution

- How far did the plane fly the first 2 h? (825 + 865 = 1,690 km)
- Which operation would you use to find out how much farther the plane has to fly? (subtraction)

TEACHING ACTIONS
8. Discuss solution(s).
9. Discuss related problems and extension.
10. Discuss special features as needed.
AFTER

Finding the Answer

Choose the Operations

$$\begin{array}{r} 825 \\ +\ 865 \\ \hline 1{,}690 \end{array} \quad \rightarrow \quad \begin{array}{r} 4{,}140 \\ -1{,}690 \\ \hline 2{,}450 \end{array}$$

The plane has to fly 2,450 km farther.

Related Problems: 58, 48, 43, 33, 23

Problem Extension

If it takes the plane 5 h to reach San Francisco, what is the average speed in kilometers per hour? (828 km/h)

79 Process Problem

Mr. Tanabe stood on the middle rung of a ladder, washing windows on an office building. He stepped up 3 rungs to reach more of the windows. Then he saw a spot he had missed on one of the windows below so he climbed down 5 rungs. After he cleaned that spot, he climbed up 7 rungs and washed the rest of the windows. After he finished, he climbed the remaining 6 rungs to the top of the building, where he stored all of his materials. How many rungs did the ladder have?

TEACHING ACTIONS

1. Read the problem.
2. Ask questions for understanding the problem.
3. Discuss possible solution strategies.

BEFORE

Understanding the Problem

- Where was Mr. Tanabe standing? (on the middle rung)
- What did he do when he saw a spot he missed? (He climbed down 5 rungs.)
- If a ladder had 3 rungs, which rung would Mr. Tanabe be standing on if he were on the middle rung? (the second)

TEACHING ACTIONS

4. Observe students.
5. Give hints as needed for solving the problem.
6. Require students to check back and answer the problem.
7. Give problem extension as needed.

DURING

Planning a Solution

- Guess how many rungs the ladder had. Try it.
- Is the number of rungs above the middle of the ladder the same as the number below the middle of the ladder? (yes)
- Draw a ladder and show were Mr. Tanabe would be standing. (See solution.)

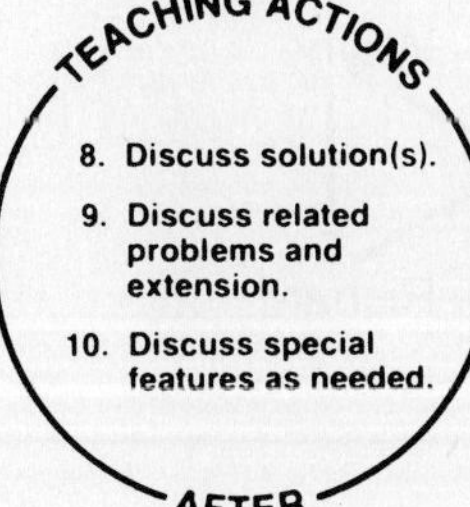

Finding the Answer

Draw a Picture

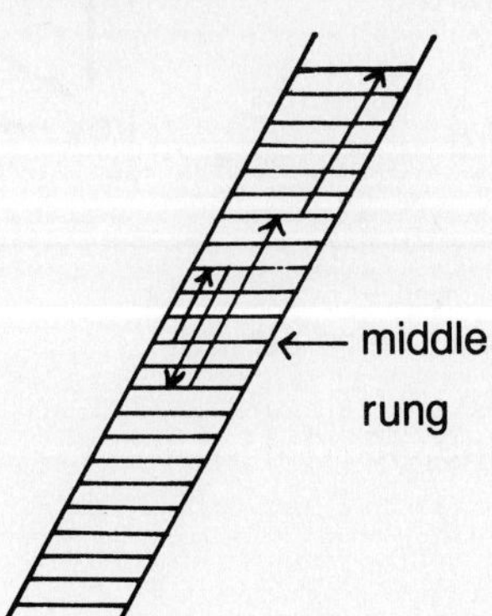

+11 rungs above the middle rung
11 rungs below the middle rung
+ 1 middle rung

23 rungs

There are 23 rungs on the ladder.

Related Problems: 45, 44, 10, 9

Problem Extension

If Mr. Tanabe started on the third rung, how many rungs would there be on the ladder? (14)

80 Process Problem

Anthony, Wayne, Troy, and Richard played checkers. Each boy played each of the other boys 1 game. How many games were played in all?

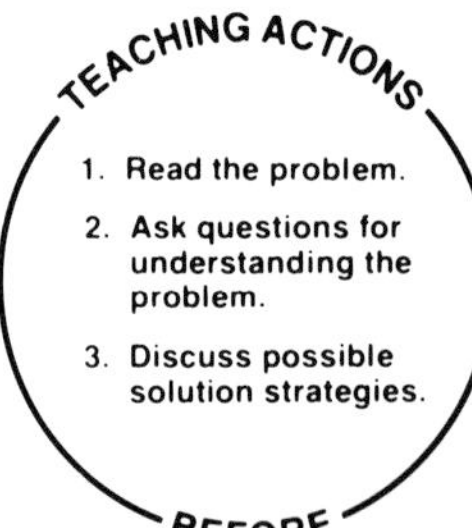

Understanding the Problem

- How many boys were in the tournament? (4)
- If Anthony played Richard, would they play each other again? (no)

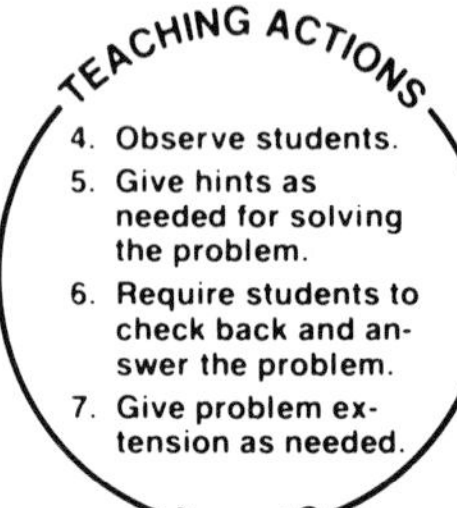

Planning a Solution

- Who did Anthony play? (Wayne, Troy, Richard)
- How many games did Anthony play? (3)
- How many *more* games did Wayne play? (2—he already played Anthony.) List them. (Troy, Richard)
- Can you list the games played? (See solution.)

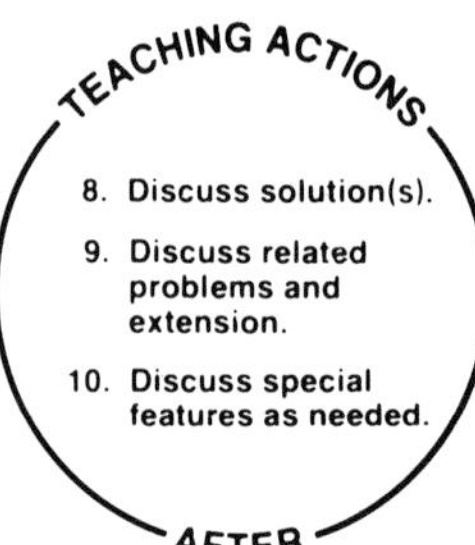

Finding the Answer

Make an Organized List

Anthony—Wayne
Anthony—Troy
Anthony—Richard

Wayne—Troy
Wayne—Richard

Troy—Richard

Draw a Picture

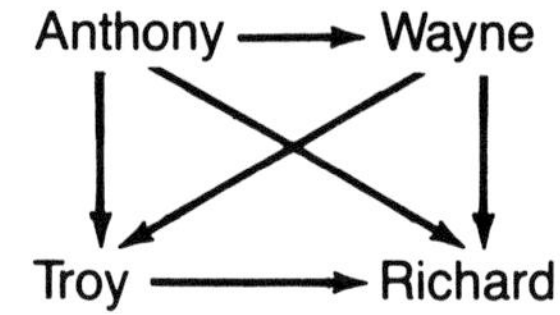

6 games of checkers were played.

Related Problems: 79, 75, 50, 49, 45

Problem Extension

Marianne and Eileen played Michael, Todd, and Wes in a checkers tournament. If each girl played each boy once, how many games were played? (6)

81 Skill Activity

Identify the operation or operations needed to solve each problem. Then choose one problem and solve it.

1. Jenny bought 14 fan magazines last year. Wendy and Janis each bought 27 fan magazines last year. How many fan magazines did the 3 girls buy altogether?

2. Heather sold 39 of her $0.12 comic books. Before she sold them she had 98 comic books. How much money did Heather make when she sold her 39 comic books?

Discussion

Possible answers:

1. Use addition and multiplication to solve this problem.

$$\begin{array}{r} 27 \\ \times\ 2 \\ \hline 54 \end{array} \qquad \begin{array}{r} 54 \\ +14 \\ \hline 68 \end{array}$$

The girls bought 68 fan magazines altogether.

2. Use multiplication to solve this problem.

$$\begin{array}{r} \$0.12 \\ \times\ 39 \\ \hline 108 \\ 36 \\ \hline \$4.68 \end{array}$$

Heather made $4.68 when she sold her comic books.

82 One-Step Problem

There were 120 students in the fourth grade last year. There were 4 classrooms. If each classroom had the same number of students, how many students were in each?

Solution

$$\begin{array}{r} 30 \\ 4\overline{)120} \\ 12 \\ \hline 0 \\ 0 \\ \hline \end{array}$$

There were 30 students in each classroom.

83 Multiple-Step Problem

One Saturday Marsha read for 25 min. Then she worked in the garden for 45 min. If she started at 11:00 a.m., what time did she finish?

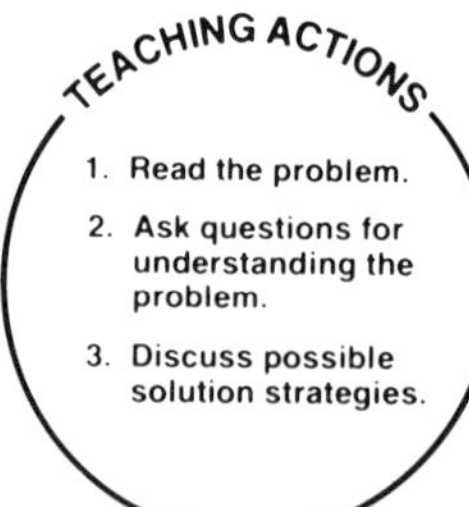

Understanding the Problem

- How many minutes did Marsha read? (25)
- How many minutes did she work in the garden? (45)
- How many minutes are there in an hour? (60)

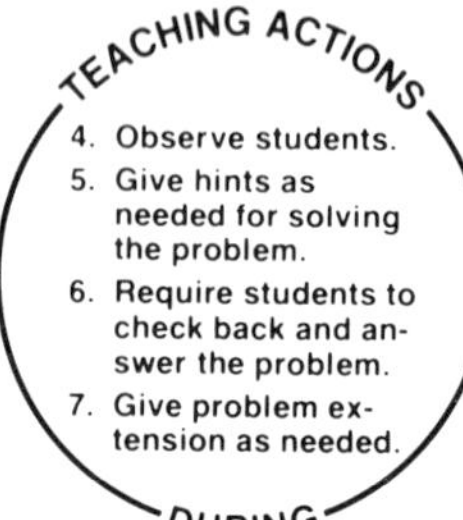

Planning a Solution

- When did Marsha begin? (11:00 a.m.)
- Which operation would you use to find out how long Marsha read and worked in the garden? (addition)

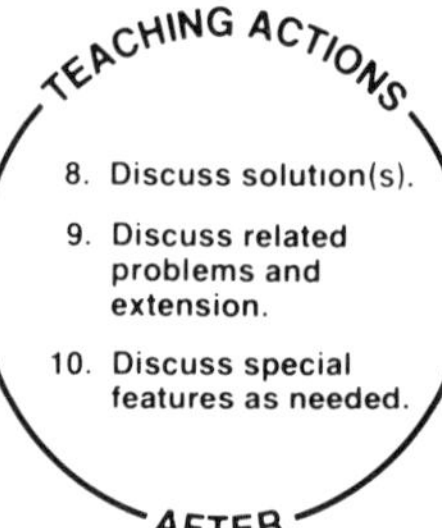

Finding the Answer

Choose the Operations

$$\begin{array}{r} 25 \\ +45 \\ \hline 70 \end{array}$$

70 (70 min = 1 h 10 min) ⟶

11:00 a.m. + 1 h 10 min = 12:10 p.m.

Marsha finished at 12:10 p.m.

Problem Extension

Suppose you know that Marsha finished at 12:30 p.m. If she read for 25 min and worked in the garden for 45 min, when did she start? (11:20 a.m.)

84 Process Problem

In Friday night's football game North High School lost to South High School 20–12. What are the different ways North High could have scored 12 points?

Touchdown—6 points
Point after touchdown—1 point
Field goal—3 points
Safety—2 points

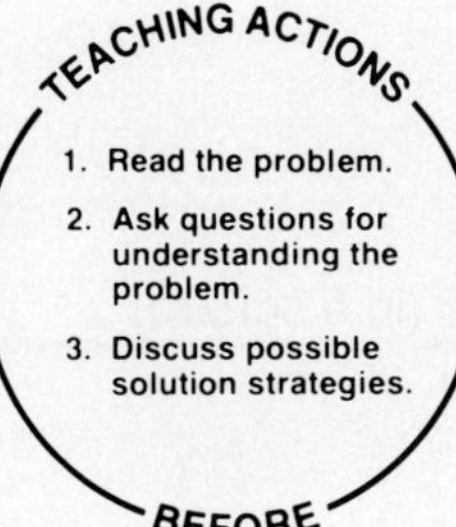

Understanding the Problem

- How many points did North High score? (12)
- Can you score a point after touchdown without scoring a touchdown? (no)
- How many points do you get for a touchdown? (6)
- Could North High have scored 3 touchdowns? (no, $6 \times 3 = 18$)

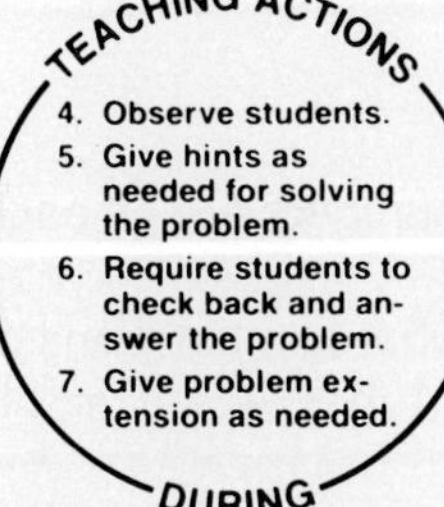

Planning a Solution

- Suppose North High scored 1 touchdown. How many points did they score? (6) How many more points are needed? (6) What is one way they can get 6 more points? (another touchdown) Are there other ways to get 6 points? (yes)
- Could North High have scored all field goals? (Yes, $3 \times 4 = 12$.)
- Try making a list of all of the ways North High could have scored 12 points. (See solution.)

Finding the Answer

Make an Organized List

Touchdown	*Point After*	*Field Goal*	*Safety*	*Points*
2	0	0	0	12
1	1	1	1	12
1	0	2	0	12
1	0	0	3	12
0	0	4	0	12
0	0	2	3	12
0	0	0	6	12

There are 7 ways North High could have scored 12 points.

Guess and Check

- Try 1 TD, 1 PAT, 2 FG—$6 + 1 + 6 = 13$. (no)
- Try 1 TD, 1 PAT, 1 FG, 1S—$6 + 1 + 3 + 2 = 12$. (yes)

and so on

Related Problems: 80, 75, 74, 50, 49

Problem Extension

How many ways could South High have scored 20 points? (16)

85 Process Problem

Juan was playing with the 4 dominoes shown below. He wanted to arrange them in a square so that each side would equal 11. How could he do this?

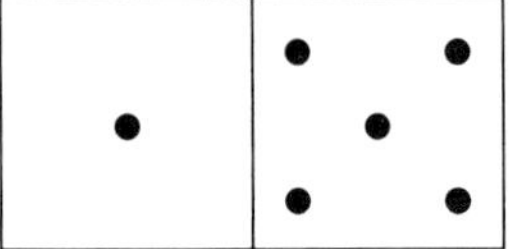
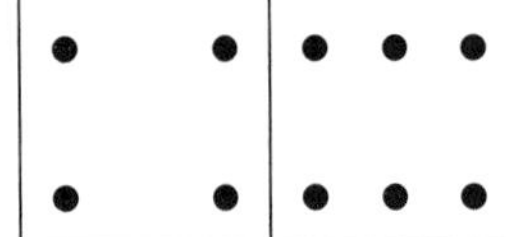
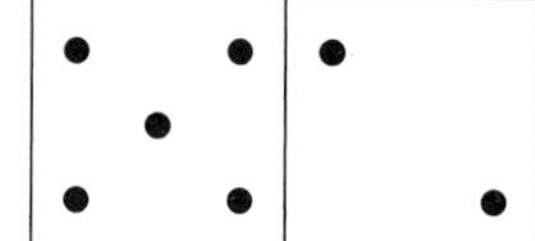
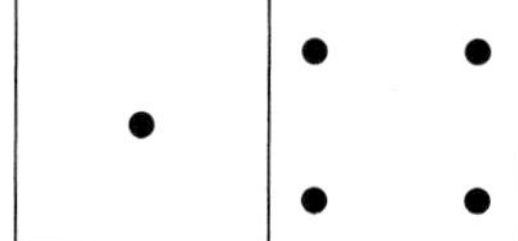

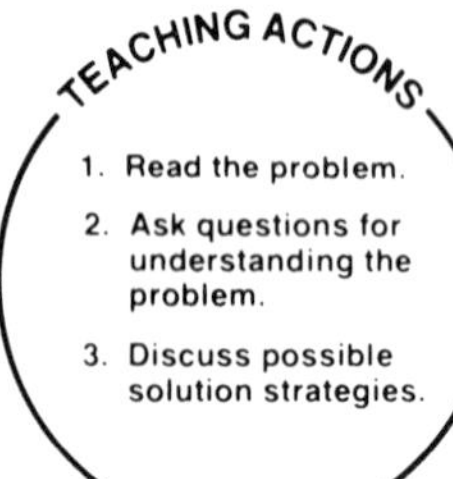

Understanding the Problem

- What is the sum for each side? (11)
- How does Juan want to arrange the dominoes? (in a square)
- How many dominoes does he have? (4)

TEACHING ACTIONS

4. Observe students.
5. Give hints as needed for solving the problem.
6. Require students to check back and answer the problem.
7. Give problem extension as needed.

DURING

Planning a Solution

- Make a square using the dominoes. How many numbers are on each side? (3)
- Using the numbers showing on the dominoes, what combinations of 3 can you think of that total 11? (5 + 1 + 5 = 11, 6 + 4 + 1 = 11, for example)

Finding the Answer

Draw a Picture

Juan arranged his dominoes so that 1, 4, and 6 were on one side; 6, 1, and 4 were on another; 4, 2, and 5 were on the third; and 5, 5, and 1 were on the fourth. All sides equaled 11.

Related Problems: 80, 79, 45, 44, 10

Problem Extension

Make a domino square using these dominoes so that all sides equal 12.

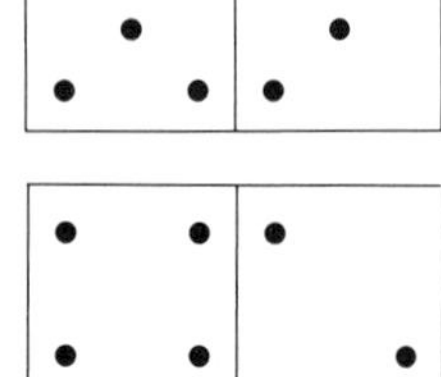
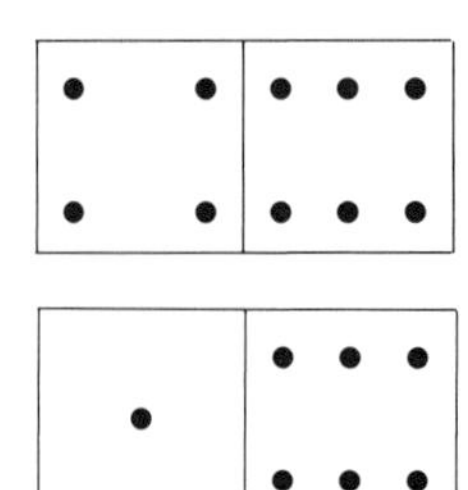
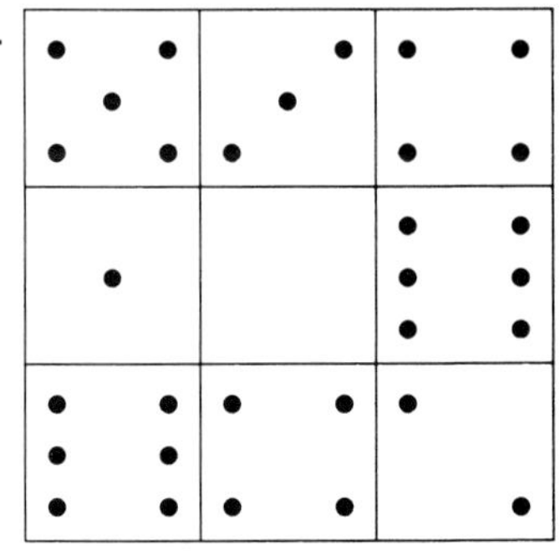

86 Skill Activity

Write a story problem that this picture would help you to solve.

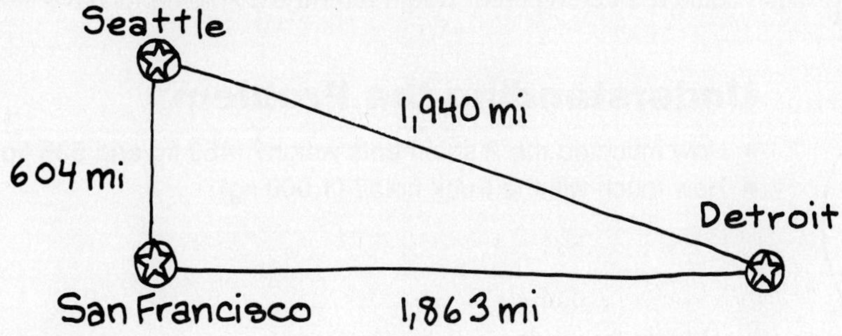

Discussion

Possible problem: Jeremy took a trip from San Francisco to Detroit, then went on to Seattle, and finally returned to San Francisco. How far did he travel?

$$\begin{array}{r} 1{,}863 \\ 1{,}940 \\ +\ 604 \\ \hline 4{,}407 \end{array}$$

He traveled 4,407 mi.

87 One-Step Problem

Chicago is 440 km from Cleveland. Charlene's dad went to Chicago on a business trip and returned to Cleveland the next day. How far did he drive in all?

Solution

$$\begin{array}{r} 440 \\ +\ 440 \\ \hline 880 \end{array} \quad \textit{or} \quad \begin{array}{r} 440 \\ \times\ \ 2 \\ \hline 880 \end{array}$$

Charlene's dad traveled 880 km.

88 Multiple-Step Problem

There are 2 shipments weighing 450 kg and 535 kg waiting at the warehouse to be shipped. Both shipments are headed for Boston. A truck that can carry a maximum of 1,000 kg arrives to pick up the 2 shipments. How much could a third shipment weigh after the 2 shipments have been loaded onto the truck?

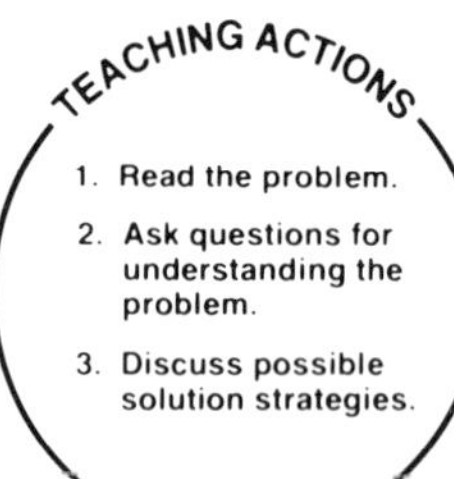

Understanding the Problem

- How much do the 2 shipments weigh? (450 kg and 535 kg)
- How much will the truck hold? (1,000 kg)

TEACHING ACTIONS

4. Observe students.
5. Give hints as needed for solving the problem.
6. Require students to check back and answer the problem.
7. Give problem extension as needed.

DURING

Planning a Solution

- What is the combined weight of the 2 shipments? (450 kg + 535 kg = 985 kg)
- Which operation would you use to find out how much more weight the truck can carry? (subtraction)

Finding the Answer

Choose the Operations

$$\begin{array}{r} 450 \\ +\underline{535} \\ 985 \end{array} \rightarrow \begin{array}{r} 1{,}000 \\ -\underline{985} \\ 15 \end{array}$$

A third shipment could weigh 15 kg.

Related Problems: 78, 58, 48, 43, 33

Problem Extension

How many 50-kg boxes can the truck carry? (20)

89 Process Problem

How many different ways can Alex make change for a 50¢ piece without using pennies?

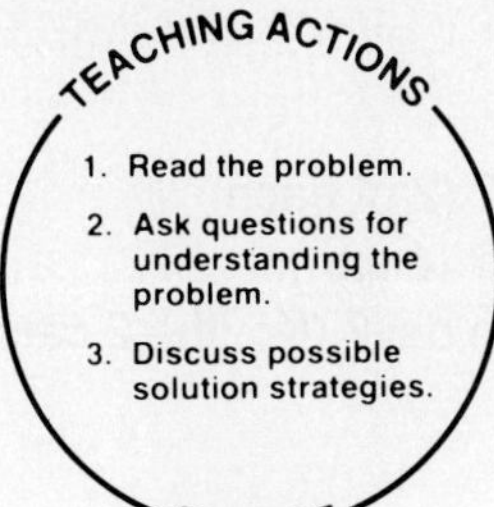

Understanding the Problem

- What coin does Alex have? (a 50¢ piece)
- Can he use pennies to make change? (no)

TEACHING ACTIONS

4. Observe students.
5. Give hints as needed for solving the problem.
6. Require students to check back and answer the problem.
7. Give problem extension as needed.

DURING

Planning a Solution

- Can Alex use a 50¢ piece to make change? (No, that's what he has.)
- What coins can he use? (quarter, dime, nickel)
- Could he make change using quarters only? (yes, 2 quarters)
- Could he use dimes only? (yes) Nickels only? (yes)
- Make a list of the ways Alex could make change. (See solution.)

TEACHING ACTIONS

8. Discuss solution(s).
9. Discuss related problems and extension.
10. Discuss special features as needed.

AFTER

Finding the Answer

Make an Organized List

N	*D*	*Q*
0	5	0
0	0	2
1	2	1
2	4	0
3	1	1
4	3	0
5	0	1
6	2	0
8	1	0
10	0	0

Guess and Check

- Try 3N + 2D + 1Q—15 + 20 + 25 = 60¢. (no)
- Try 3N + 1D + 1Q—15 + 10 + 25 = 50¢. (yes)

and so on

There are 10 possible ways that Alex can make change for a 50¢ piece without using pennies.

Related Problems: 84, 80, 75, 74, 50

Problem Extension

How many ways could Alex make change for a $1.00 bill if he used at least 1 quarter? (22)

90 Process Problem

Jeanette has 2 dogs and 2 cats. Each dog eats 1 can of dog food a day. The cats share 1 can of cat food each day. How many cans of dog and cat food do Jeanette's pets eat each week?

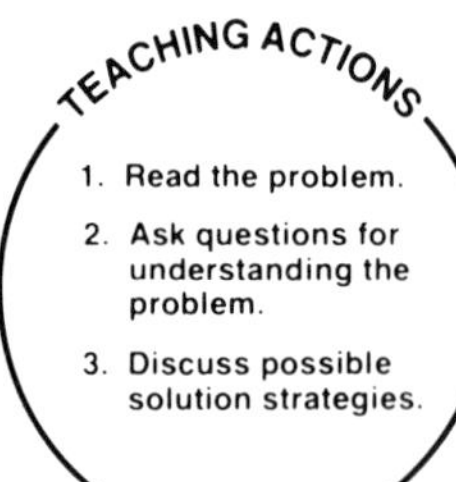

Understanding the Problem

- How many dogs and cats does Jeanette have? (2 of each)
- How much food does each dog eat in one day? (1 can)
- Does each cat eat a whole can of cat food each day? (No, the 2 cats share 1 can of cat food.)

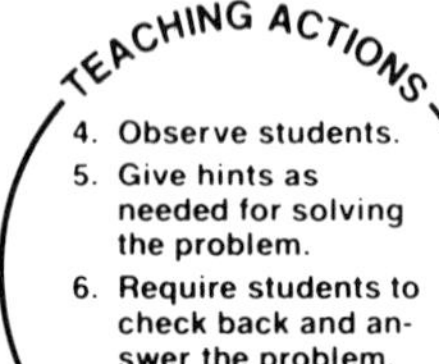

Planning a Solution

- If each dog eats 1 can of food per day, how many cans do the 2 eat in 1 day? (2) 2 days? (4)
- If the 2 cats share a can of food each day, how much will they eat in 2 days? (2 cans) 3 days? (3 cans)
- Make a table showing the days of the week and the cans of food eaten each day. (See solution.)

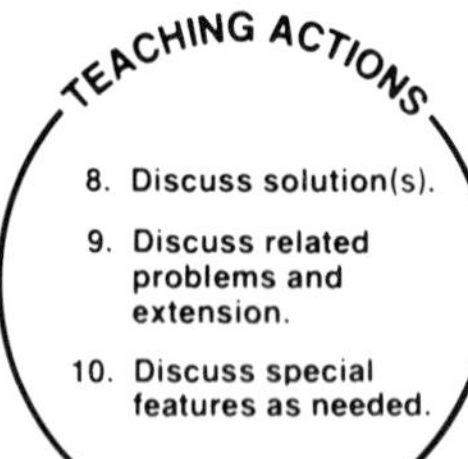

Finding the Answer

Make a Table

Day	*Cans of Dog Food*	*Cans of Cat Food*
1	2	1
2	2	1
3	2	1
4	2	1
5	2	1
6	2	1
7	2	1

14 + 7 = 21

Jeanette's dogs and cats eat 21 cans of food each week.

Related Problems: 65, 64, 55, 54, 40

Problem Extension

If dog food costs $0.49 per can and cat food costs $0.35 per can, how much does Jeanette spend per week for pet food? ($9.31)

91 Skill Activity

Write a question that can be answered by using the data in this story. Then find the answer.

There are 7 red stripes and 6 white stripes on the American flag. Ms. Michael has 4 flags.

Discussion

Possible question:

What is the total number of stripes on the flags?

$$\begin{array}{r} 7 \\ \times\ 4 \\ \hline 28 \end{array} \qquad \begin{array}{r} 6 \\ \times\ 4 \\ \hline 24 \end{array} \qquad \begin{array}{r} 28 \\ +24 \\ \hline 52 \end{array}$$

The total number of stripes on Ms. Michael's flags is 52.

92 One-Step Problem

Mercury is 58 million km from the sun. Earth is 299 million km from the sun, and Pluto is 5,879 million km from the sun. How much farther is Earth from the sun than Mercury is from the sun?

Solution

$$\begin{array}{r} 299 \\ -\ 58 \\ \hline 241 \end{array}$$

Earth is 241 million km farther from the sun than Mercury.

Note: The distance of Pluto from the sun is unnecessary data.

93 Multiple-Step Problem

The members of the Garcia family have taken 4 equal servings of corn flakes. Each has a 35-gram (g) serving. How many grams of corn flakes are left in the 335-g box?

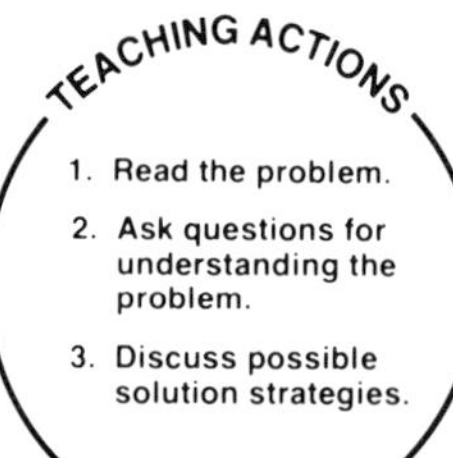

Understanding the Problem

- How many servings were taken? (4)
- How much was each serving? (35 g)
- What was the total weight of the corn flakes box? (335 g)

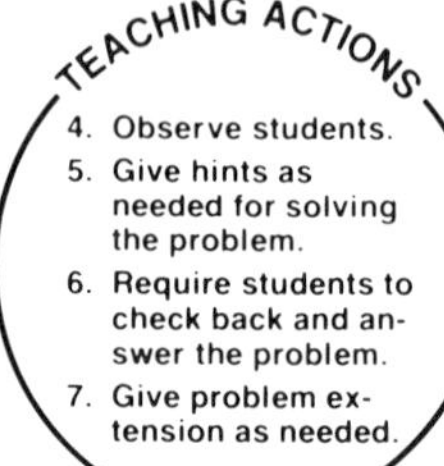

Planning a Solution

- If 1 serving weighs 35 g, how much do 4 servings weigh? (35 g × 4 = 140 g)
- Which operation would you use to find out how many grams are left in the box? (subtraction)

Finding the Answer

Choose the Operations

$$\begin{array}{r} 35 \\ \times\ 4 \\ \hline 140 \end{array} \rightarrow \begin{array}{r} 335 \\ -140 \\ \hline 195 \end{array}$$

There are 195 g of corn flakes left in the box.

Related Problems: 73, 63, 38, 28

Problem Extension

Suppose Mr. Garcia ate twice as much as everyone else. How many grams are left in the box? (160)

94 Process Problem

The town of Marble Farms lies on Poplar Lake. The town of Crazy Knob is west of Marble Farms. Daisy Grove is east of Crazy Knob but west of Marble Farms. Winding Knoll is east of Rolling Ridge but west of Daisy Grove and Crazy Knob. Which town is farthest west?

TEACHING ACTIONS

1. Read the problem.
2. Ask questions for understanding the problem.
3. Discuss possible solution strategies.

BEFORE

Understanding the Problem

- Where is Marble Farms located? (on Poplar Lake)
- Where is Crazy Knob? (west of Marble Farms)
- Where is Daisy Grove? (east of Crazy Knob, west of Marble Farms)
- Where is Winding Knoll? (east of Rolling Ridge, west of Daisy Grove and Crazy Knob)

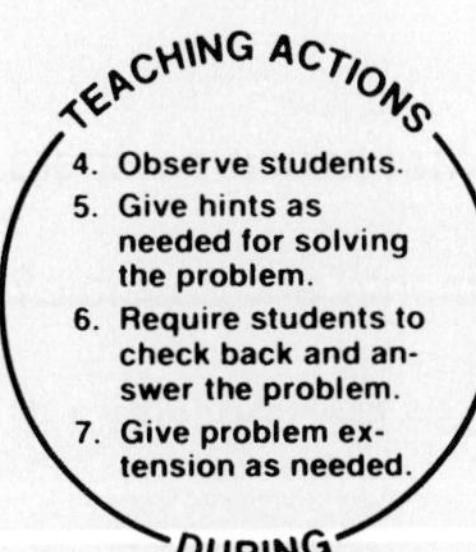

Planning a Solution

- Is Marble Farms the farthest west? (No, Daisy Grove and Crazy Knob are west of it.)
- Try drawing a picture and show Marble Farms and Crazy Knob. (See solution.)

TEACHING ACTIONS

8. Discuss solution(s).
9. Discuss related problems and extension.
10. Discuss special features as needed.

AFTER

Finding the Answer

Draw a Picture

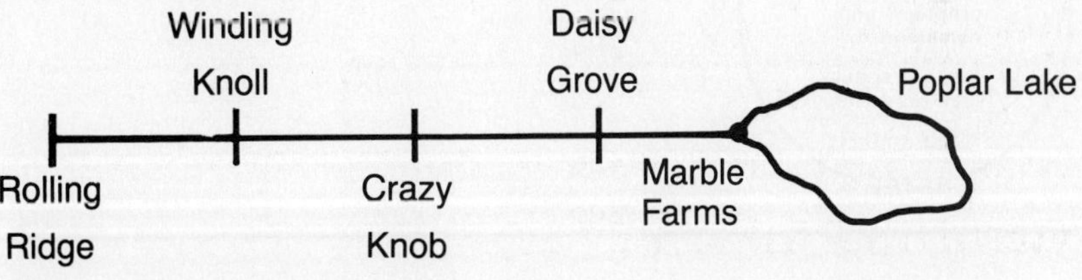

Rolling Ridge is farthest west.

Related Problems: 85, 80, 79, 45, 44

Problem Extension

Suppose Daisy Grove is west of Winding Knoll but east of Rolling Ridge. Crazy Knob is west of Rolling Ridge but east of Marble Farms. Which town is farthest east? (Winding Knoll)

95 Process Problem

There were 3 wooden numbers—2, 4, and 6—for sale at a garage sale. Marcy bought them to put on her house to show her house number. What 3-digit house numbers could Marcy make from these 3 wooden numbers?

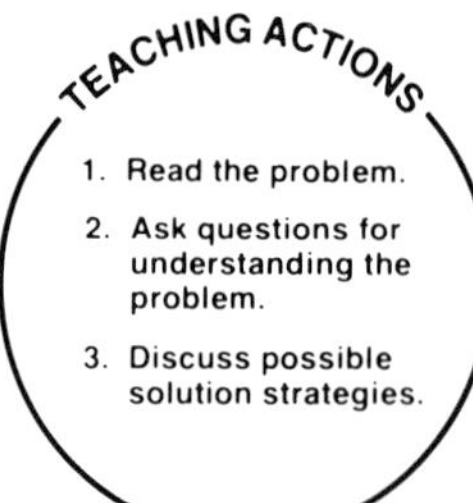

Understanding the Problem

- How many numbers does Marcy have? (3) What are they? (2, 4, 6)
- Does she want to use all 3 numbers? (yes)

TEACHING ACTIONS

4. Observe students.
5. Give hints as needed for solving the problem.
6. Require students to check back and answer the problem.
7. Give problem extension as needed.

DURING

Planning a Solution

- Suppose she had only 2 numbers—2 and 4. What house numbers could she make? (24 and 42)
- If the 2 is the first of 3 numbers, what arrangements are possible? (246, 264)
- List all the numbers starting with a 2, then all those starting with a 4, then all those starting with a 6. (See solution.)

TEACHING ACTIONS

8. Discuss solution(s).
9. Discuss related problems and extension.
10. Discuss special features as needed.

AFTER

Finding the Answer

Make an Organized List

246
264

426
462

624
642

Marcy could make 6 different house numbers—
246, 264, 426, 462, 624, and 642.

Related Problems: 89, 84, 80, 75, 50

Problem Extension

If Marcy had 4 numbers—1, 2, 3, and 4—how many different 3-digit house numbers could she make? (24)

96 Skill Activity

Write a story problem that this organized list would help you to solve.

vanilla—vanilla
vanilla—chocolate
vanilla—strawberry
vanilla—walnut

chocolate—chocolate
chocolate—strawberry
chocolate—walnut

strawberry—strawberry
strawberry—walnut

walnut—walnut

Discussion

Possible problem:

Ahmad ordered 2 scoops of ice cream for dessert. If he had a selection of vanilla, chocolate, strawberry, and walnut, how many choices would he have? (Ahmad would have 10 different choices.)

97 One-Step Problem

The round-trip distance between Paris and Vienna is 147 km. Last week, Rose drove to Paris and back to Vienna 6 times. How far did she drive?

Solution

$$\begin{array}{r} 147 \\ \times \underline{\quad 6} \\ 882 \end{array}$$

Rose drove 882 km last week.

98 Multiple-Step Problem

Matthew pays \$0.18 for a newspaper and sells it for \$0.25. If he sells 20 papers each day, how much does he earn in 1 day?

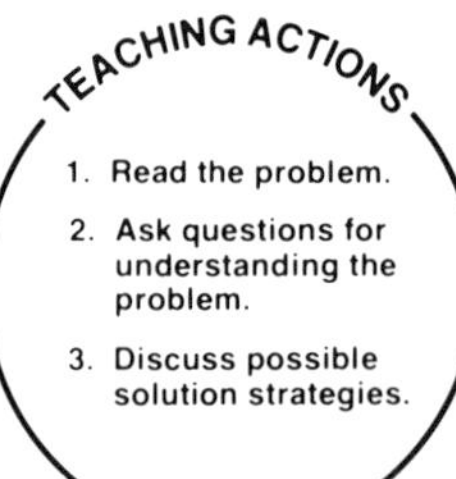

Understanding the Problem

- How much does Matthew pay for each paper? (\$0.18)
- For how much does Matthew sell each paper? (\$0.25)
- How many papers does he sell each day? (20)

TEACHING ACTIONS

4. Observe students.
5. Give hints as needed for solving the problem.
6. Require students to check back and answer the problem.
7. Give problem extension as needed.

DURING

Planning a Solution

- How much money does Matthew earn from 1 newspaper? (\$0.25 − \$0.18 = \$0.07)
- How can you find his earnings for 20 newspapers? (Multiply the amount he earns for 1 paper by 20.)

TEACHING ACTIONS

8. Discuss solution(s).
9. Discuss related problems and extension.
10. Discuss special features as needed.

AFTER

Finding the Answer

Choose the Operations

$$\begin{array}{r} \$0.25 \\ -\ 0.18 \\ \hline \$0.07 \end{array} \rightarrow \begin{array}{r} \$0.07 \\ \times\ \ 20 \\ \hline \$1.40 \end{array}$$

Matthew earns \$1.40 in 1 day.

Related Problems: 93, 73, 63, 38, 28

Problem Extension

Matthew buys the Sunday paper for \$0.50 and sells it for \$0.75. If he delivers his other newspapers 6 days and the Sunday paper on Sunday, how much does he earn each week? (\$13.40)

99 Process Problem

Tanya and Lucy were throwing balls into these baskets at the fair. Whoever scored exactly 34 points with the least number of throws won the prize. What is the least number of throws needed to score exactly 34?

TEACHING ACTIONS — BEFORE

1. Read the problem.
2. Ask questions for understanding the problem.
3. Discuss possible solution strategies.

Understanding the Problem

- What score won the prize? (34)
- What are the different numbers showing on the baskets? (2, 6, 9)

TEACHING ACTIONS — DURING

4. Observe students.
5. Give hints as needed for solving the problem.
6. Require students to check back and answer the problem.
7. Give problem extension as needed.

Planning a Solution

- Suppose you threw 2 balls. What is the highest score you can make? ($2 \times 9 = 18$)
- Can you throw all 9s and get 34? (No, because 34 is not a multiple of 9.)
- Try listing all the ways you can make 34. (See solution.)

TEACHING ACTIONS — AFTER

8. Discuss solution(s).
9. Discuss related problems and extension.
10. Discuss special features as needed.

Finding the Answer

Make an Organized List

2	*6*	*9*	
17	0	0	
14	1	0	
11	2	0	
8	3	0	
8	0	2	
5	4	0	
2	2	2	**(least)**

Guess and Check

- Try 4 nines = 36 (no)
- Try 2 nines, 0 sixes, 8 twos = 34 (yes, 10 throws)

and so on

The least number of throws needed to score 34 is 6.

Related Problems: 95, 89, 84, 80, 75

Problem Extension

What is the least number of throws needed to score 100? (13)

100 Process Problem

Nancy had 69¢ in coins. Yoshi asked her for change for a half dollar. Nancy tried to make change but found that she didn't have the correct coins to do so. What coins did she have if each of her coins was less than a half dollar?

TEACHING ACTIONS

1. Read the problem.
2. Ask questions for understanding the problem.
3. Discuss possible solution strategies.

BEFORE

Understanding the Problem

- How much money did Nancy have? (69¢)
- What did Yoshi need? (change for a half dollar)
- Can Nancy make change for a half dollar? (no)

TEACHING ACTIONS

4. Observe students.
5. Give hints as needed for solving the problem.
6. Require students to check back and answer the problem.
7. Give problem extension as needed.

DURING

Planning a Solution

- What coins could Nancy have? (pennies, nickels, dimes, quarters)
- If Nancy had 2 quarters in her coins, could she make change for a half dollar? (yes)
- Make a list of the different coins Nancy might have. (See solution.)

Finding the Answer

Make an Organized List

Pennies	Nickels	Dimes	Quarters	69¢	Change for a Half Dollar?
9	0	1	2	x	yes
9	2	0	2	x	yes
9	1	3	1	x	yes
4	1	1	2	x	yes
4	3	0	2	x	yes
4	0	4	1	x	no

Nancy had 4 pennies, 4 dimes, and 1 quarter.

Related Problems: 99, 95, 89, 84, 80

Problem Extension

How many different ways can you make change for 35¢ if you cannot use pennies? (6)

101 Skill Activity

These problems have the numbers omitted. Identify the operation or operations needed to solve each problem.

1. Rita ran a different number of miles on Monday, Tuesday, Wednesday, and Thursday. How far did she run?

2. Last week Tim bought a can of tennis balls. Tennis balls were on sale this week, so he bought another can. How much money did he save by buying the can on sale?

Discussion

Possible answers:

1. Use addition to solve this problem.

2. Use subtraction to solve this problem.

102 One-Step Problem

Use this table to find the average number of points scored by a player on the Dragon's basketball team.

Heidi	**7**
Crescenda	**12**
Ellen	**11**
Deana	**6**

Solution

$$\begin{array}{r} 7 \\ 12 \\ 11 \\ +\ 6 \\ \hline 36 \end{array} \qquad 4\overline{)36}^{\,9}$$

The average number of points scored is 9.

103 Multiple-Step Problem

Felipe and Luis sold blackberries. Felipe sold 2 gallons (gal) for $2.00 each, and Luis sold 3 gal for $1.75 each. How much money did they make together?

TEACHING ACTIONS — BEFORE

1. Read the problem.
2. Ask questions for understanding the problem.
3. Discuss possible solution strategies.

Understanding the Problem

- How many gallons of berries did Felipe sell? (2) How much did he charge for each? ($2.00)
- How many gallons of berries did Luis sell? (3) How much did he charge for each? ($1.75)

TEACHING ACTIONS — DURING

4. Observe students.
5. Give hints as needed for solving the problem.
6. Require students to check back and answer the problem.
7. Give problem extension as needed.

Planning a Solution

- If Felipe sold 1 gal for $2.00, how much did he get for 2 gal? ($4.00)
- If Luis sold 1 gal for $1.75, how much did he get for 3 gal? ($5.25)
- Which operation would you use to find out how much they made together? (addition)

TEACHING ACTIONS — AFTER

8. Discuss solution(s).
9. Discuss related problems and extension.
10. Discuss special features as needed.

Finding the Answer

Choose the Operations

$$\begin{array}{r}\$2.00\\ \times\ 2\\ \hline \$4.00\end{array} \rightarrow \begin{array}{r}\$1.75\\ \times\ 3\\ \hline \$5.25\end{array} \rightarrow \begin{array}{r}\$5.25\\ +\ 4.00\\ \hline \$9.25\end{array}$$

Together, Felipe and Luis made $9.25.

Related Problems: 68, 53, 8

Problem Extension

Each boy saved $1.00 of his earnings, and then they each went to the movies and paid $2.00 to get in. How much did each have left to spend? (Felipe had $1.00 left to spend, and Luis had $2.25 left to spend.)

104 Process Problem

Gary went to the grocery store and bought apples and bananas. The apples cost $0.10 each and the bananas cost $0.12 each. He spent exactly $1.00. How many of each kind did he buy?

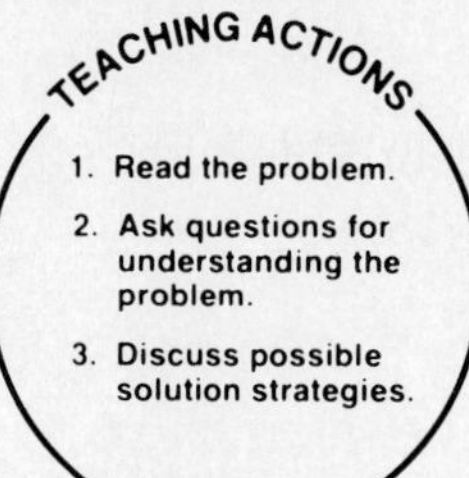

Understanding the Problem

- How much does each apple cost? ($0.10)
- How much does each banana cost? ($0.12)
- How much did Gary spend? (exactly $1.00)

TEACHING ACTIONS

4. Observe students.
5. Give hints as needed for solving the problem.
6. Require students to check back and answer the problem.
7. Give problem extension as needed.

DURING

Planning a Solution

- Can Gary buy 10 apples? (No, he has to buy some of each kind.)
- If he buys 1 apple and 1 banana, how much does he spend? ($0.10 + $0.12 = $0.22)
- Make a table to help you.

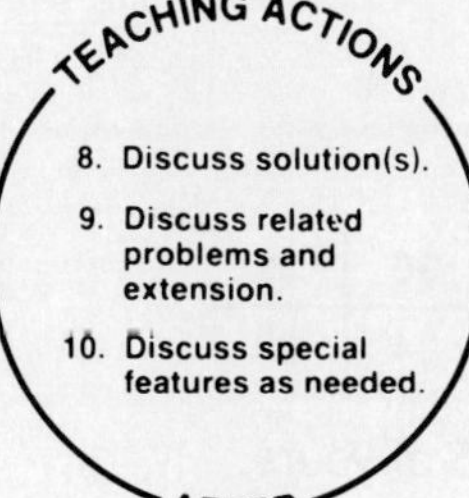

Finding the Answer

Make a Table

Bananas		*Apples*		
Number	*Cost*	*Number*	*Cost*	*Total Cost*
8	$0.96	1	$0.10	$0.96 + $0.10 = $1.06 (no)
7	0.84	1	0.10	0.84 + 0.10 = 0.94 (no)
6	0.72	2	0.20	0.72 + 0.20 = 0.92 (no)
7	0.84	2	0.20	0.84 + 0.20 = 1.06 (no)
5	0.60	4	0.40	0.60 + 0.40 = 1.00 (yes)

Guess and Check

- Try 2 bananas, 8 apples—$0.12 × 2 = $0.24, $0.10 × 8 = $0.80, $0.24 + $0.80 = $1.04. (no)
  and so on

Gary bought 5 bananas and 4 apples.

Related Problems: 99, 90, 89, 84, 75

Problem Extension

If bananas cost $0.14 each and apples cost $0.12 each, how many of each could he buy? (3 bananas and 4 apples)

105 Process Problem

A secondhand store will trade 4 of their comic books for 5 of yours. How many of their comic books will they trade for 35 of yours?

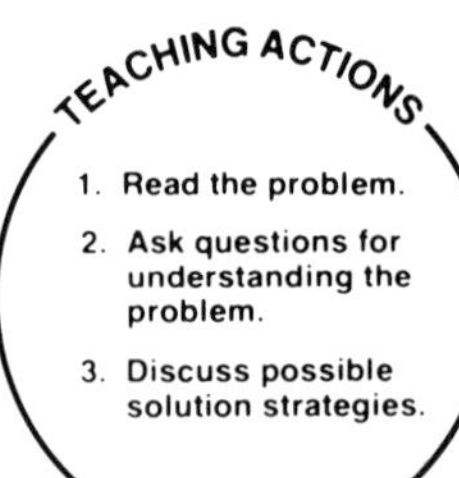

Understanding the Problem

- How many comic books will the store trade you for 5 of yours? (4)
- How many comic books do you have to trade? (35)

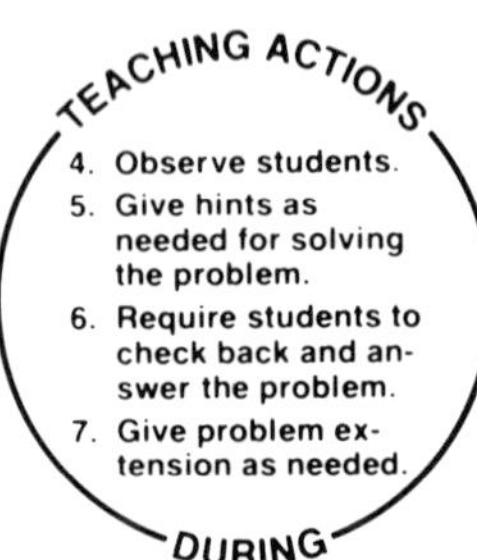

Planning a Solution

- If you gave the store 5 comic books, how many would you get? (4) 10 comic books? (8)
- Make a table to show how many comic books the store would give you for 35 of yours. (See solution.)

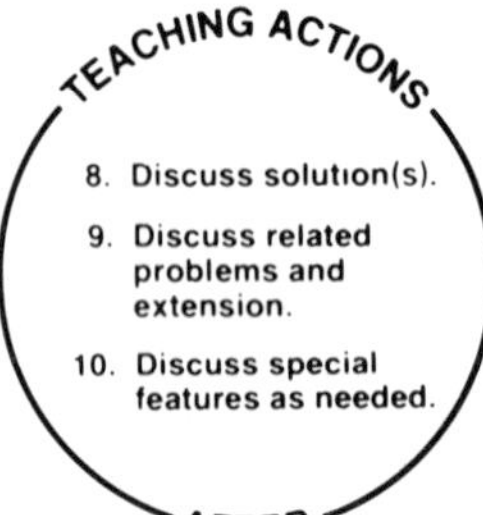

Finding the Answer

Make a Table

Yours	5	10	15	20	25	30	35
Theirs	4	8	12	16	20	24	28

The store will trade 28 of their comic books for 35 of yours.

Related Problems: 104, 90, 65, 64, 55

Problem Extension

If they gave you 40 comic books, how many did you give them? (50)

106 Skill Activity

Determine whether the answers given for these problems are reasonable. If an answer is not reasonable, explain why.

1. Brad's brother works 40 hours each week and earns $8.50 an hour. How much money does Brad's brother earn each week?
   Answer: Brad's brother earns about $350 each week.

2. Jack has 144 empty bottles, and a case holds 24 bottles. How many cases does Jack need to hold the 144 bottles?
   Answer: Jack needs 10 cases.

Discussion

1. This answer is reasonable because $40 \times 8 = \$320$ and $40 \times 9 = \$360$.

2. This answer is not reasonable because $10 \times 24 = 240$ and Jack has only 144 bottles. The correct answer is 6 cases.

107 One-Step Problem

How much cloth does Jonah need to cover the tabletop?

4 ft
6 ft

Solution

$$\begin{array}{r} 4 \\ \times\ 6 \\ \hline 24 \end{array}$$

Jonah needs 24 sq ft of cloth.

108 Multiple-Step Problem

Mr. Kenner drives to and from work each day. His house is 15 km from his office. How many kilometers does he drive if he works 5 days a week?

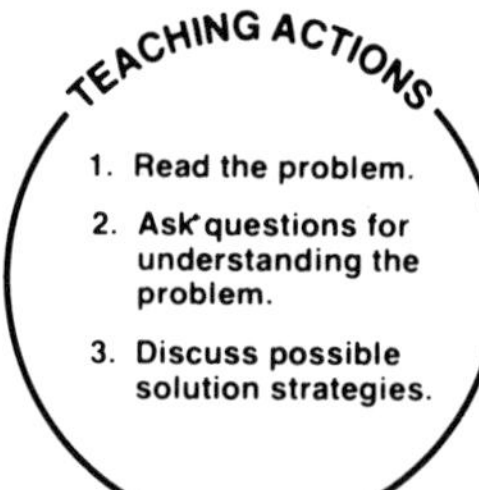

Understanding the Problem

- How many days does Mr. Kenner work? (5)
- What distance is 15 km? (the distance from his house to his office)

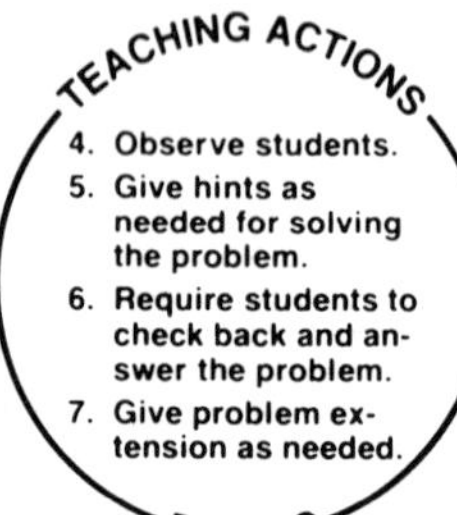

Planning a Solution

- If Mr. Kenner drives to work and then home, does he drive 15 km? (No, he drives 30 km round trip.)
- If you know the round-trip distance for 1 day, which operation would you use to find the distance for 5 days? (multiplication)

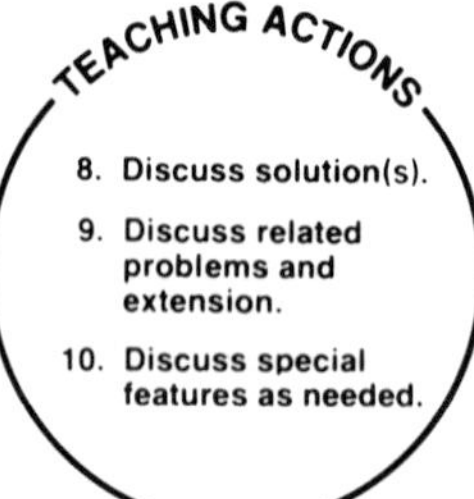

Finding the Answer

Choose the Operations

$$\begin{array}{r} 15 \\ \times\ 2 \\ \hline 30 \end{array} \rightarrow \begin{array}{r} 30 \\ \times\ 5 \\ \hline 150 \end{array}$$

Mr. Kenner drives 150 km per week.

Problem Extension

If Mr. Kenner's car gets about 15 km/L of gasoline, about how many liters of gasoline does he use each week to drive to and from work? (10)

109 Process Problem

Conrad, Charles, and Eugene live next door to each other. One of them is an engineer, one is a teacher, and one is a lawyer. Find each man's occupation from these clues: 1. Charles lives in the middle apartment. 2. When Eugene goes away, his dog is fed by the teacher. 3. The engineer knocks on Conrad's wall when his stereo is too loud.

TEACHING ACTIONS
1. Read the problem.
2. Ask questions for understanding the problem.
3. Discuss possible solution strategies.
BEFORE

Understanding the Problem

- What are the men's names? (Conrad, Charles, Eugene)
- What are their occupations? (engineer, teacher, lawyer)
- Do you know each man's occupation? (No, that is what we are trying to find out.)
- Where does Charles live? (in the middle apartment)
- Which man knocks on Conrad's wall when the stereo is too loud? (the engineer)

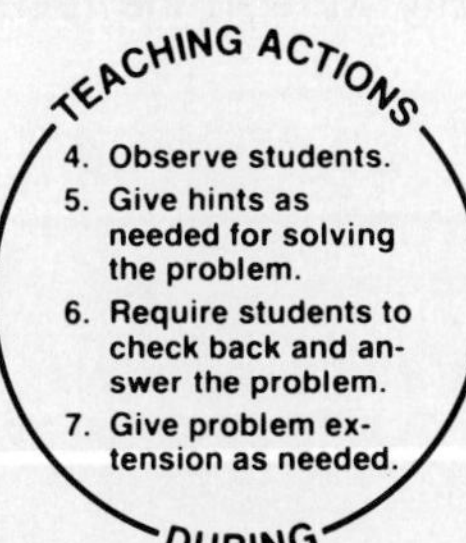

Planning a Solution

- If Charles lives in the middle apartment, where must Conrad and Eugene live? (in the end or outside apartments)
- If Conrad lives in the end apartment, who is the only person who could knock on his wall? (Charles) What is Charles' occupation? (engineer)
- If Eugene is not the teacher, then who must the teacher be? (Conrad)
- Try drawing and labeling a picture. (See solution.)

TEACHING ACTIONS
8. Discuss solution(s).
9. Discuss related problems and extension.
10. Discuss special features as needed.
AFTER

Finding the Answer

Draw a Picture

Knocks on wall — Not a teacher

Conrad (teacher)

Charles (engineer)

Eugene (lawyer)

Use Logical Reasoning

Conrad—end—teacher (feeds Eugene's dog)

Charles—middle—engineer (knocks on Conrad's wall)

Eugene—end—lawyer

Conrad is the teacher, Charles is the engineer, and Eugene is the lawyer.

Related Problems: 94, 85, 80, 79, 74

Problem Extension

Pat, Billy, John, and Alexander ran a race. Alexander was ahead of John, but behind Billy. Pat was behind John. Who won the race? (Billy)

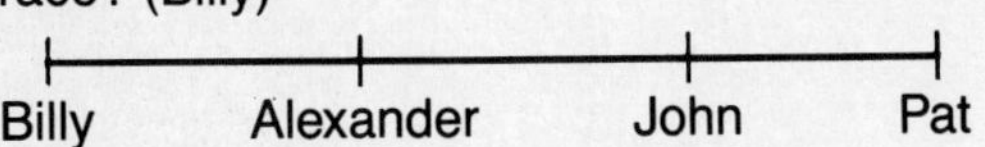

110 Process Problem

Five girls ran the 50-m race. Angela came in first. Betty came in last. If Darla was ahead of Claudia and Ellen was just behind her, who came in second?

TEACHING ACTIONS
1. Read the problem.
2. Ask questions for understanding the problem.
3. Discuss possible solution strategies.

BEFORE

Understanding the Problem

- How many girls ran the race? (five)
- Who came in first? (Angela)
- Who came in last? (Betty)
- Who was ahead of Claudia? (Darla)
- Who was just behind her? (Ellen)

TEACHING ACTIONS
4. Observe students.
5. Give hints as needed for solving the problem.
6. Require students to check back and answer the problem.
7. Give problem extension as needed.

DURING

Planning a Solution

- Draw a picture to show where Angela and Betty were in the race. (See solution.)

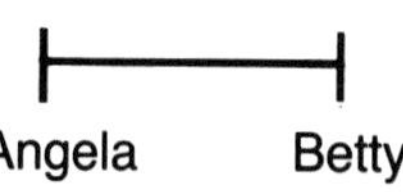

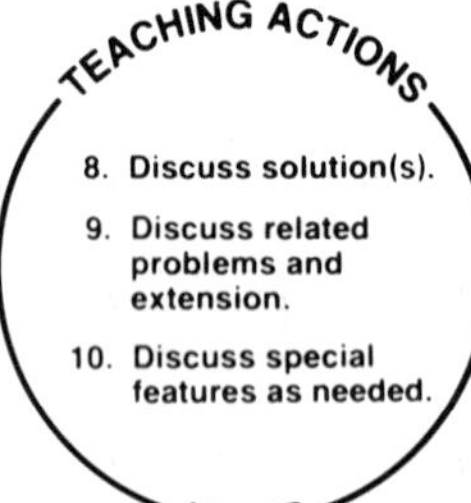

Finding the Answer

Draw a Picture

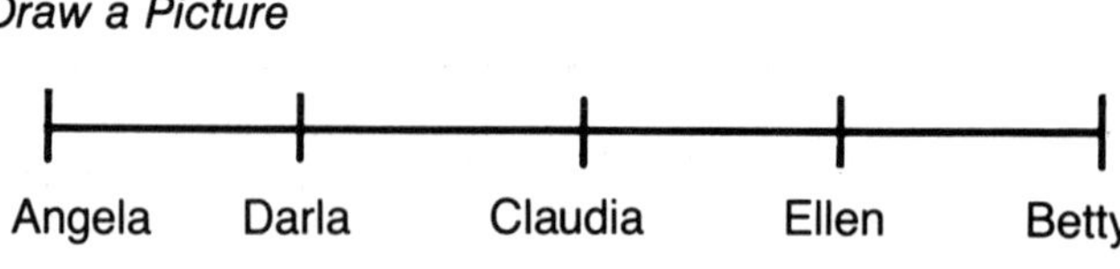

Darla came in second.

Related Problems: 109, 94, 85, 80, 79

Problem Extension

There are two possible solutions to the problem above, depending on whom the word *her* is referring to. What is the other possible solution?

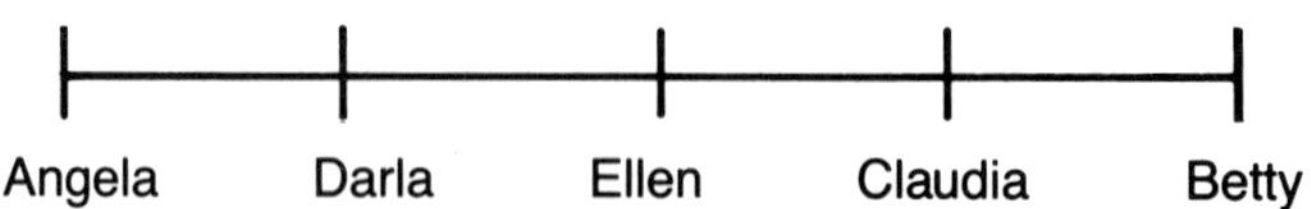

Ellen is either just behind Darla or just behind Claudia.

111 Skill Activity

Write a story problem that this picture would help you to solve.

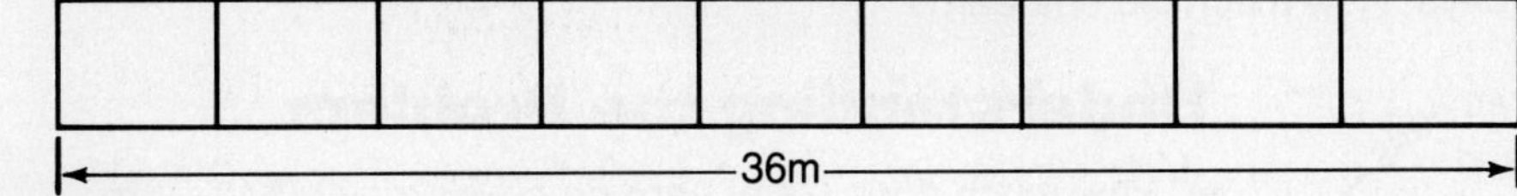

Discussion

Possible problem:

Max had a pipe that was 36 m long. He cut the pipe into 9 equal parts. How long was each part?

$$9\overline{)36}\ \text{with quotient } 4$$

Each part was 4 m long.

112 One-Step Problem

Cheng and his sister each made \$13.20 in the walkathon. Laurence and his sister each made \$19.80 in the walkathon. How much more did Laurence make than Cheng's sister?

Solution

$$\begin{array}{r} \$19.80 \\ -\ \underline{13.20} \\ \$\ 6.60 \end{array}$$

Laurence made \$6.60 more than Cheng's sister.

113 Multiple-Step Problem

Carol earns $4 a day on weekdays and $6 a day on Saturdays. Last month she worked 13 weekdays and 3 Saturdays. How much did she earn?

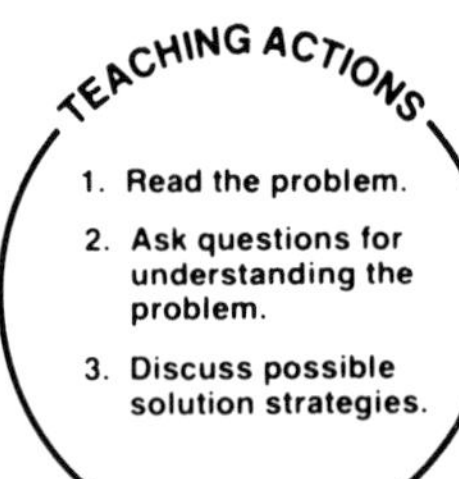

Understanding the Problem

- How much does Carol earn on a weekday? ($4)
- How much does she earn on Saturdays? ($6)
- How many weekdays did she work last month? (13) Saturdays? (3)

TEACHING ACTIONS
4. Observe students.
5. Give hints as needed for solving the problem.
6. Require students to check back and answer the problem.
7. Give problem extension as needed.
DURING

Planning a Solution

- If she makes $4 a day for each weekday and she worked 13 weekdays, how much did she earn? (13 × $4 = $52)
- If she makes $6 for each Saturday, how much will she earn for 3 Saturdays? (3 × $6 = $18)
- Which operation would you use to find out how much she earned last month? (addition)

Finding the Answer

Choose the Operations

13		3		$52.00
×$ 4	→	×$ 6	→	+ 18.00
$52		$18		$70.00

Carol earned $70 last month.

Related Problems: 103, 68, 53, 8

Problem Extension

Suppose Carol spent $10.00 for gas, $9.75 for meals, and $8.00 for parking. What was her profit? ($42.25)

114 Process Problem

Zelda was arranging coins in the following manner: 1 penny, followed by 2 nickels, followed by 3 dimes, followed by 4 pennies, and so on. If she continues in this manner, what coin will be in the thirtieth position?

TEACHING ACTIONS

1. Read the problem.
2. Ask questions for understanding the problem.
3. Discuss possible solution strategies.

BEFORE

Understanding the Problem

- What coin comes first? (1 penny) Followed by? (2 nickels) Then? (3 dimes)

TEACHING ACTIONS

4. Observe students.
5. Give hints as needed for solving the problem.
6. Require students to check back and answer the problem.
7. Give problem extension as needed.

DURING

Planning a Solution

- If we have 1 penny, 2 nickels, and 3 dimes, how many coins do we have in all? (6)
- Which coin is in the fifth position? (dime)
- Could you draw a picture and look for a pattern? (See solution.)

TEACHING ACTIONS

8. Discuss solution(s).
9. Discuss related problems and extension.
10. Discuss special features as needed.

AFTER

Finding the Answer

Draw a Picture

PNNDDDPPPPNNNNNDDDDDDPPPPPPPN(N)NNNNNN

Make a Table/Look for a Pattern

Coin	P	N	D	P	N	D	P	N	D
Number of Coins	1	2	3	4	5	6	7	8	9
Total Coins	1	3	6	10	15	21	28→	36	45

A nickel is in the thirtieth position.

Related Problems: 105, 104, 90, 65, 64

Problem Extension

Suppose we had 1 penny. When we added our nickels we had 4 coins. We added our dimes and had 9 coins. What coin would be in the thirtieth position? (a dime)

115 Process Problem

Jerry was mowing his lawn when he noticed that Christy was also mowing her lawn next door. They stopped to talk, and Jerry learned that Christy mows her lawn once every 8 days. Jerry mows his lawn once every 6 days. In how many days will they be mowing their lawns together again?

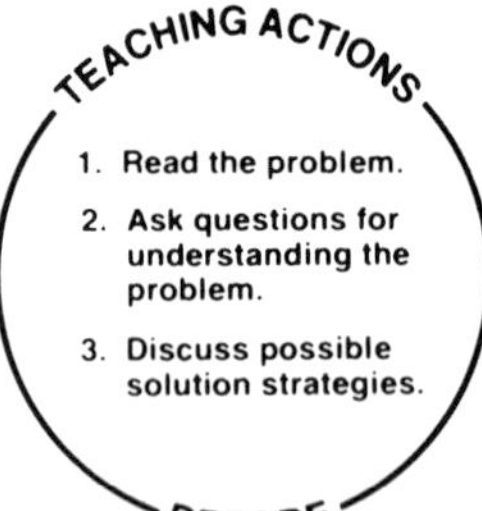

Understanding the Problem

- How often does Christy mow her lawn? (every 8 days)
- How often does Jerry mow his lawn? (every 6 days)

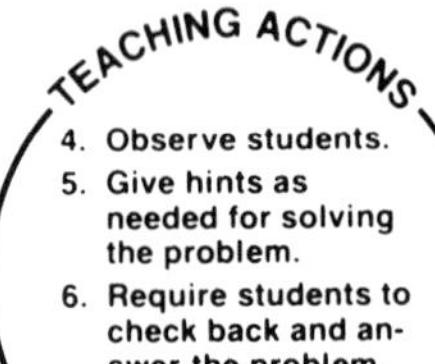

Planning a Solution

- In how many days will Christy mow her lawn again? (8) Then the next time? (16)
- When will Jerry mow his lawn again? (in 6 days) Then again? (12)
- Try making a table. (See solution.)

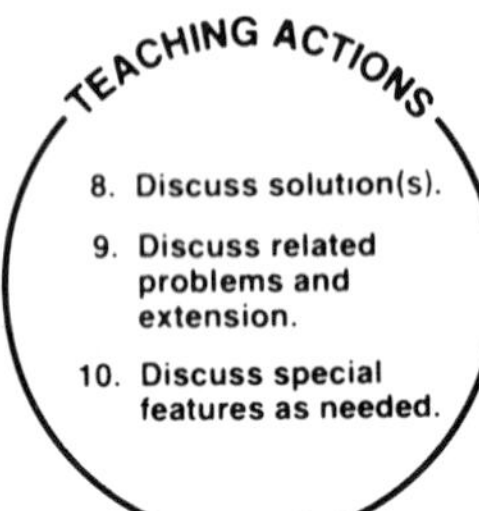

Finding the Answer

Make a Table

Christy	8	16	(24)	32
Jerry	6	12	18	(24)

Jerry and Christy will be mowing their lawns together again in 24 days.

Related Problems: 114, 105, 104, 90, 65

Problem Extension

Suppose Jerry's mom pays him $1.50 each time he mows the lawn and Christy's mom pays her $1.75 each time she mows the lawn. Who will have made the most money in 24 days? (Jerry)

116 Skill Activity

Here is a problem and its solution. Write the answer in a complete sentence.

A jet can hold 250 people, and 135 people have already made reservations. If 125 more people make reservations, will the plane be overbooked? If so, by how many people? Answer: Yes, 10

Discussion

Possible answer:

The plane will be overbooked by 10 people.

117 One-Step Problem

Jennifer needs 40 bottles of juice for her party next week. She goes to the store and finds that juice is sold in cartons of 6. How many cartons of juice will she have to buy?

Solution

$$\begin{array}{r} 6 \quad \text{R4} \\ 6\overline{)40} \\ \underline{36} \\ 4 \end{array}$$

Jennifer will have to buy 7 cartons of juice.

118 Multiple-Step Problem

Dwayne and 7 friends each sold the same number of tickets to the fun fair. They sold 56 tickets in all. How many tickets were sold by each person?

TEACHING ACTIONS

1. Read the problem.
2. Ask questions for understanding the problem.
3. Discuss possible solution strategies.

BEFORE

Understanding the Problem

- How many tickets were sold in all? (56)
- Who sold the tickets? (Dwayne and 7 friends)
- Did they each sell the same number? (yes)

TEACHING ACTIONS

4. Observe students.
5. Give hints as needed for solving the problem.
6. Require students to check back and answer the problem.
7. Give problem extension as needed.

DURING

Planning a Solution

- How many people sold tickets? ($1 + 7 = 8$)
- Which operation would you use to find out how many tickets were sold by each person? (division)

TEACHING ACTIONS

8. Discuss solution(s).
9. Discuss related problems and extension.
10. Discuss special features as needed.

AFTER

Finding the Answer

Choose the Operations

$$\begin{array}{r} 1 \\ +7 \\ \hline 8 \end{array} \rightarrow \begin{array}{r} 7 \\ 8\overline{)56} \end{array}$$

Each person sold 7 tickets.

Problem Extension

If the tickets sold for \$2.00 each and the boys made \$0.50 on each ticket sold, how much was their total profit? (\$28.00)

119 Process Problem

Chris, Ned, Terry, and Jack were seeing who could throw the baseball the farthest. Chris threw the ball 10 m farther than Jack. Jack threw it 7 m farther than Terry. Terry threw it 9 m less than Ned. Ned threw the ball 45 m. How far did Chris throw the ball?

Understanding the Problem

- How far did Ned throw the ball? (45 m)
- How far did Terry throw the ball? (9 m less than Ned)
- How far did Jack throw the ball? (7 m farther than Terry)
- How far did Chris throw the ball? (10 m farther than Jack)

TEACHING ACTIONS
4. Observe students.
5. Give hints as needed for solving the problem.
6. Require students to check back and answer the problem.
7. Give problem extension as needed.
DURING

Planning a Solution

- If Ned threw the ball 45 m and Terry threw it 9 m less, how far did Terry throw the ball? (45 m − 9m = 34 m)
- How far did Jack throw the ball? (34 m + 7 m = 41 m)
- Begin with the distance Ned threw the ball and work backwards. (See solution.)

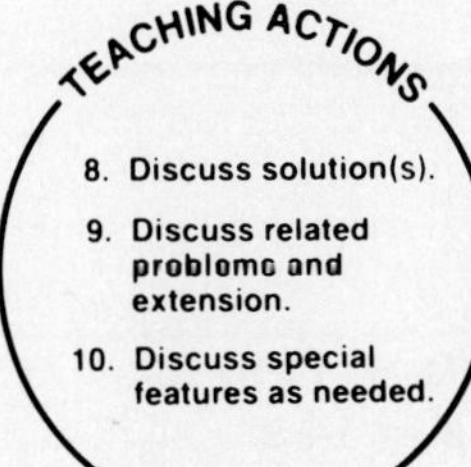

Finding the Answer

Work Backwards

Ned threw the ball 45 m.
Terry threw 9 m less than Ned—45 m − 9 m = 34 m.
Jack threw 7 m farther than Terry—34 m + 7 m = 41 m.
Chris threw 10 m farther than Jack—41 m + 10 m = 51 m.

Chris threw the ball 51 m.

Related Problems: 60, 59, 25, 24

Problem Extension

Suppose Terry threw the ball 10 m farther than Ned. How far would each boy have thrown? (Ned, 45 m; Terry, 55 m; Jack, 62 m; Chris, 72 m)

120 Process Problem

Mr. Sivertson divided his class into 5 teams. They were having relay races. Team A finished 5 s before Team B. Team B finished 2 s before Team C. Team C finished 3 s after Team D. Team D finished 7 s before Team E. Team E finished in 18 s. How many seconds did it take for Team A to finish?

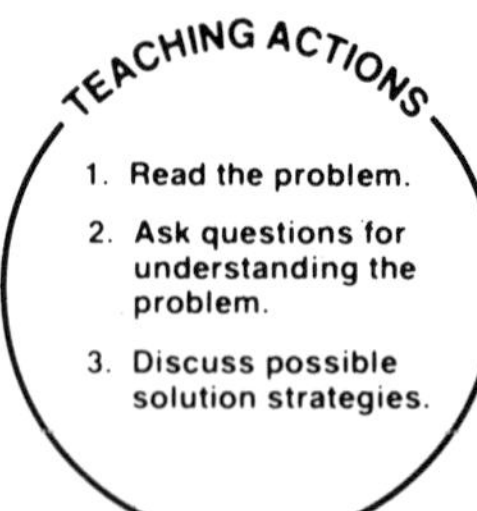

Understanding the Problem

- How long did it take for Team E to finish? (18 s)
- When did Team D finish? (7 s before Team E) Team C? (3 s after Team D) Team B? (2 s before Team C) Team A? (5 s before Team B)

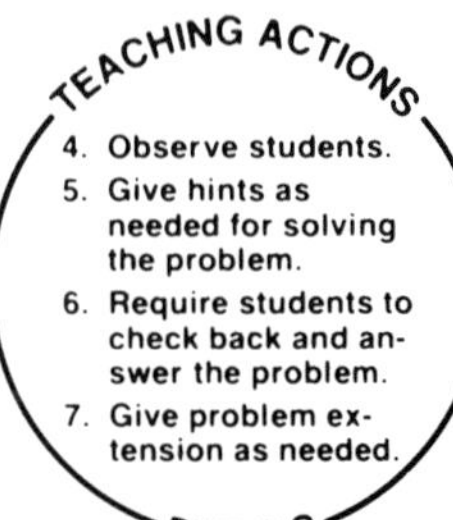

Planning a Solution

- If Team E finished in 18 s and Team D finished 7 s before them, when did Team D finish? (in 11 s)
- Begin with Team E and work backwards. (See solution.)

Finding the Answer

Work Backwards

Team E finished in 18 s.
Team D finished 7 s before Team E—18 s − 7s = 11 s.
Team C finished 3 s after Team D—11 s + 3 s = 14 s.
Team B finished 2 s before Team C—14 s − 2s = 12 s.
↓ Team A finished 5 s before Team B—12 s − 5 s = 7 s.

Team A finished in 7 s.

Related Problems: 119, 60, 59, 25, 24

Problem Extension

Suppose Team D finished 4 s after Team A. When would each team finish? (Team E, 18 s; D, 22 s; C, 25 s; B, 23 s; A, 16 s)

121 Skill Activity

Write a story problem that you can solve by using the following number sentence:

$$18 + 23 + 40 = ?$$

Discussion

Possible problem:

Dorothy had 18 books, Gloria had 23 books, and Judy had 40 books. How many books did they have in all?

122 One-Step Problem

Willie's coach got 10 new baseball bats for his Little League team. There were 25 children on the team. The bats cost $5.95 each. How much did Willie's coach pay for the bats?

Solution

```
  $5.95
×    10
-------
 $59.50
```

Willie's coach paid $59.50 for the bats.

Note: The fact that there were 25 children on the team is unnecessary data.

123 Multiple-Step Problem

Ms. Jarzinski went on a vacation. She drove 80 km the first day and stopped to visit her sister. Then she drove 75 km to visit her brother. Finally she drove 95 km back to her home. If her car gets 10 km per liter (L) of gasoline, how many liters of gasoline did she use on her trip?

TEACHING ACTIONS

1. Read the problem.
2. Ask questions for understanding the problem.
3. Discuss possible solution strategies.

BEFORE

Understanding the Problem

- How far did Ms. Jarzinski drive to her sister's? (80 km) To her brother's? (75 km) Back home? (95 km)
- How many km/L does her car get? (10)

TEACHING ACTIONS

4. Observe students.
5. Give hints as needed for solving the problem.
6. Require students to check back and answer the problem.
7. Give problem extension as needed.

DURING

Planning a Solution

- How long was her trip? (80 km + 75 km + 95 km = 250 km)
- Which operation would you use to find out how much gasoline she used? (division)

TEACHING ACTIONS

8. Discuss solution(s).
9. Discuss related problems and extension.
10. Discuss special features as needed.

AFTER

Finding the Answer

Choose the Operations

```
  80            25
  75   →   10)250
+ 75           20
----           --
 250           50
               50
               --
```

Ms. Jarzinski used 25 L of gasoline on her trip.

Related Problem: 118

Problem Extension

If gasoline costs \$0.50/L, how much did Ms. Jarzinski spend on gasoline? (\$12.50)

124 Process Problem

Eric had $24 of his birthday money to spend. How many fish can Eric buy with his money?

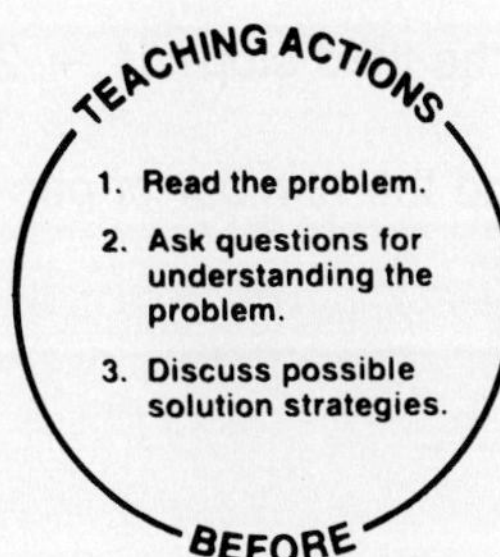

Understanding the Problem

- What was Joe's sale? (4 fish for $6, get 1 free)
- How much money did Eric have? ($24)

TEACHING ACTIONS

4. Observe students.
5. Give hints as needed for solving the problem.
6. Require students to check back and answer the problem.
7. Give problem extension as needed.

DURING

Planning a Solution

- If you paid $6 for 4 fish and got 1 free, how many fish did you get for $6? (4 + 1 = 5)
- How many fish would you get for $12? (10)
- Try making a table. (See solution.)

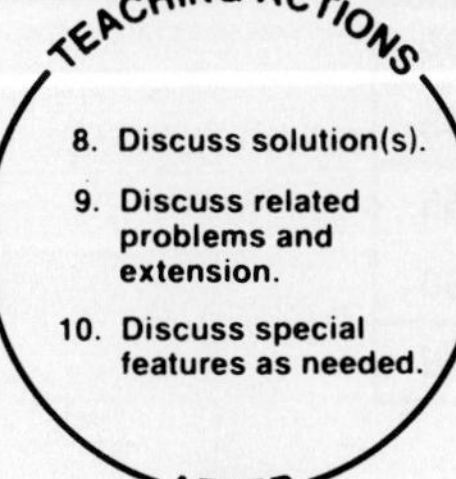

Finding the Answer

Make a Table

(2 free if you buy 8) ↗

	4 + 1	8 + 2	12 + 3	16 + 4
Number of fish	5	10	15	20
Cost	$6	$12	$18	$24

Eric could get 20 fish for $24.

Related Problems: 115, 114, 105, 104, 90

Problem Extension

Juanita jogs 7 m a day and Isabel jogs 9 m a day. How many miles will Juanita have jogged when Isabel has jogged 63 m? (49)

125 Process Problem

A train will hold 78 passengers. If it starts out empty and picks up 1 passenger at the first stop, 2 passengers at the second stop, 3 passengers at the third stop, and so on, after how many stops will it be full?

TEACHING ACTIONS

1. Read the problem.
2. Ask questions for understanding the problem.
3. Discuss possible solution strategies.

BEFORE

TEACHING ACTIONS

4. Observe students.
5. Give hints as needed for solving the problem.
6. Require students to check back and answer the problem.
7. Give problem extension as needed.

DURING

TEACHING ACTIONS

8. Discuss solution(s).
9. Discuss related problems and extension.
10. Discuss special features as needed.

AFTER

Understanding the Problem

- How many passengers can the train hold? (78)
- How many passengers get on the train at the first stop? (1) The second stop? (2) The third stop? (3)

Planning a Solution

- How many passengers are on the train after the second stop? (1 + 2 = 3)
- How many passengers are on the train after the third stop? (1 + 2 + 3 = 6)
- Make a table showing the number of stops and the number of passengers on the train. (See solution.)

Finding the Answer

Make a Table/Look for a Pattern

Stop	*Number of Passengers Getting On*	*Total Passengers on Train*
1	1	1
2	2	1 + 2 = 3
3	3	1 + 2 + 3 = 6
4	4	1 + 2 + 3 + 4 = 10
5	5	1 + 2 + 3 + 4 + 5 = 15
6	6	1 + 2 + 3 + . . . + 6 = 21
7	7	1 + 2 + 3 + . . . + 7 = 28
8	8	1 + 2 + 3 + . . . + 8 = 36
9	9	1 + 2 + 3 + . . . + 9 = 45
10	10	1+ 2 + 3 + . . . + 10 = 55
11	11	1+ 2 + 3 + . . . + 11 = 66
12	12	1+ 2 + 3 + . . . + 12 = 78

Pattern: The number of passengers getting on increases by 1 at each stop.

The train will be full after the twelfth stop.

Related Problems: 124, 115, 114, 105, 104

Problem Extension

Suppose the train picks up 2 passengers at the first stop, 4 passengers at the second stop, 6 passengers at the third stop, and so on. (Remember that the train holds 78 passengers.) When will it be full? (At the eighth stop the train would have 62 passengers, and 18 would be waiting at the ninth stop, so only 16 of them could get on.)

126 Skill Activity

Write a story problem that you can solve by using the following number sentence:

$$20 \times 8 = 160$$

Discussion

Possible problem:

There were 20 cases of bottles in the store. Each case had 8 bottles in it. How many bottles were there in all?

127 One-Step Problem

A door has a height of 7 feet. How many inches tall is the door?

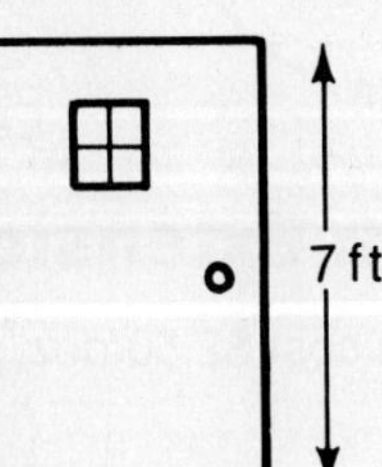

Solution

$$\begin{array}{r} 7 \\ \times 12 \\ \hline 84 \end{array}$$

The door is 84 inches tall.

128 Multiple-Step Problem

Ms. Gomez purchased 4 dozen (doz) doughnuts for the 16 women in her sewing group. All the doughnuts were eaten. How many doughnuts did each woman eat if each ate the same number?

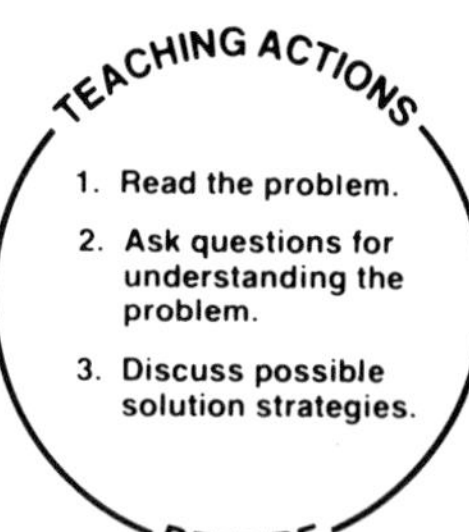

Understanding the Problem

- How many doughnuts did Ms. Gomez purchase? (4 doz) How many are there in 1 doz? (12)
- How many women were there in the sewing group? (16)
- Did they all eat the same number of doughnuts? (yes)

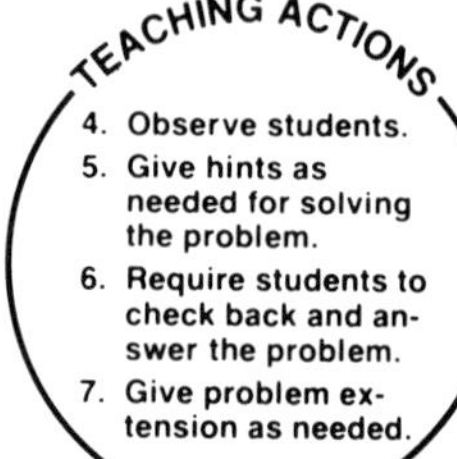

Planning a Solution

- If there are 12 doughnuts in 1 doz, how many are there in 4 doz? ($12 \times 4 = 48$)
- Which operation would you use to find out how many each woman ate? (division)

Finding the Answer

Choose the Operations

$$\begin{array}{r} 12 \\ \times\ 4 \\ \hline 48 \end{array} \rightarrow 16\overline{)48}\ \text{quotient } 3,\quad \begin{array}{r} 48 \\ \hline \end{array}$$

Each woman ate 3 doughnuts.

Problem Extension

If doughnuts cost $2.50/doz, how much did Ms. Gomez spend for the doughnuts? ($10.00)

129 Process Problem

Mr. Lee is a car salesman. Last month he sold 3 used cars for every 4 new cars. He sold a total of 56 cars. How many new cars did he sell?

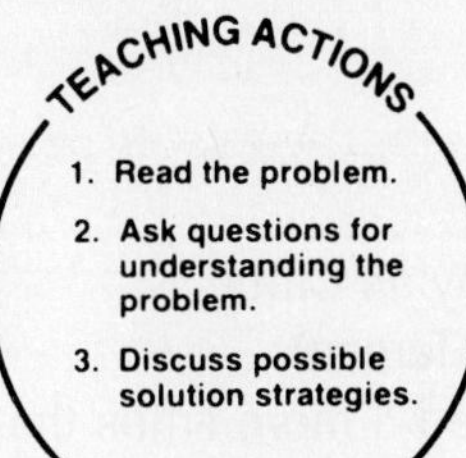

Understanding the Problem

- How many cars did he sell altogether? (56)
- What was the ratio of used cars sold to new cars sold? (3 used for every 4 new)

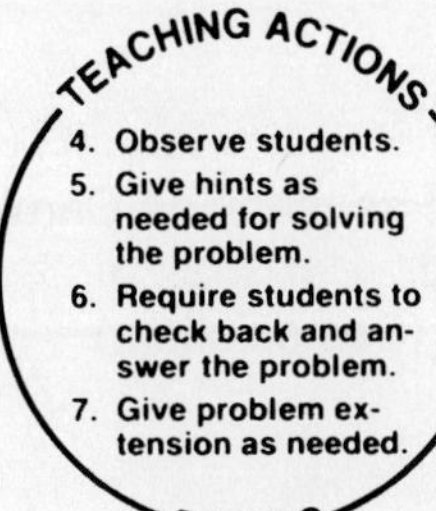

Planning a Solution

- If he sold 6 used cars, how many new cars did he sell? (8) For a total of how many? (14)
- Try making a table. (See solution.)

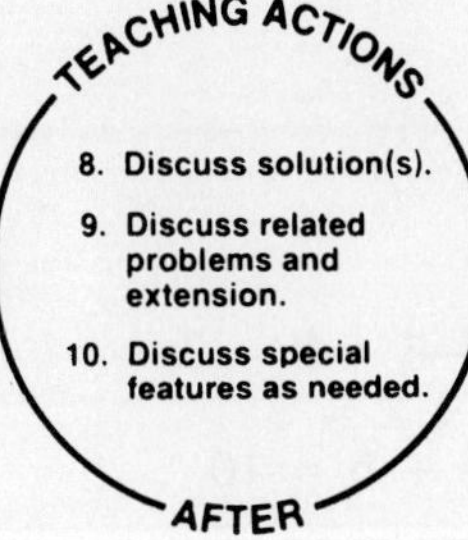

Finding the Answer

Make a Table

Used	*New*	*Total*
3	4	7
6	8	14
9	12	21
12	16	28
15	20	35
18	24	42
21	28	49
24	32	56

Guess and Check

- Try 15 used + 20 new = 35. (no)
- Try 21 used + 28 new = 49. (no)

and so on

When Mr. Lee sold 56 cars, he had sold 32 new cars.

Related Problems: 125, 124, 115, 114, 105

Problem Extension

If he sold 48 new cars, how many used cars did he sell? (36)

130 Process Problem

Melanie, Teresa, Carol, and Janet were skipping rope. Melanie skipped 5 more times than Janet. Janet skipped 2 more times than Teresa. Teresa skipped half as many times as Carol. Carol skipped 6 times. How many times did Melanie skip?

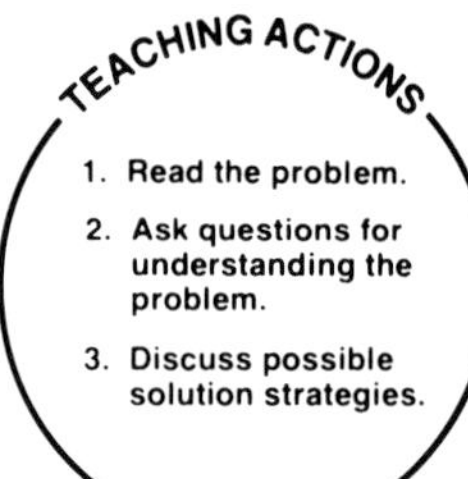

Understanding the Problem

- How many times did Carol skip? (6)
- How many times did Teresa skip? (half as many as Carol)
- How many times did Janet skip? (2 more than Teresa)
- What do we know about Melanie? (She skipped 5 more times than Janet.)

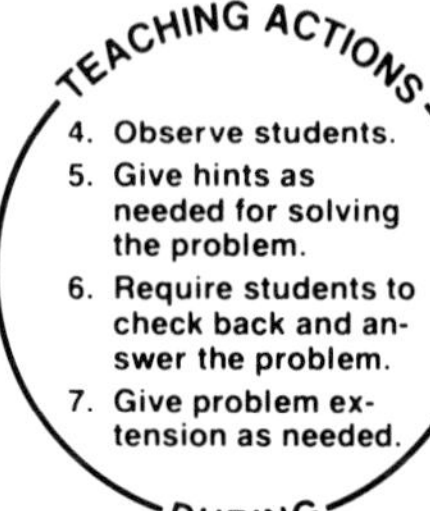

Planning a Solution

- If Carol skipped 6 times and Teresa skipped half as many as Carol, how many times did Teresa skip? (half of 6—3)
- Begin with Carol's number and work backwards. (See solution.)

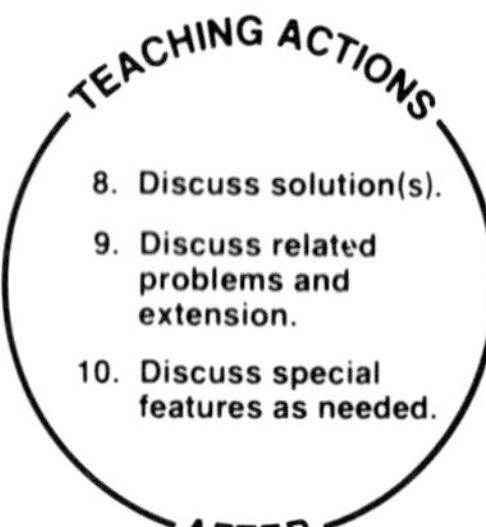

Finding the Answer

Work Backwards

Carol skipped 6 times.
Teresa skipped half as many times as Carol—$6 \div 2 = 3$.
Janet skipped 2 more times than Teresa—$3 + 2 = 5$.
↓ Melanie skipped 5 more times than Janet—$5 + 5 = 10$.

Melanie skipped 10 times.

Related Problems: 120, 119, 60, 59, 25

Problem Extension

Suppose Teresa skipped twice as many times as Carol. How many times would each girl have skipped? (Carol, 6; Teresa, 12; Janet, 14; Melanie, 19)

131 Skill Activity

This problem contains unnecessary data. Select the data you need to solve the problem. Then find the answer.

The diameter of the earth is about 8,000 miles. The diameter of the moon is about 2,200 miles, and the diameter of the sun is about 864,000 miles. How much greater is the earth's diameter than the moon's diameter?

Discussion

You don't need to know the sun's diameter.

$$\begin{array}{r} 8{,}000 \\ -\,2{,}200 \\ \hline 5{,}800 \end{array}$$

The earth's diameter is 5,800 miles greater than the moon's.

132 One-Step Problem

Find the average morning temperature.

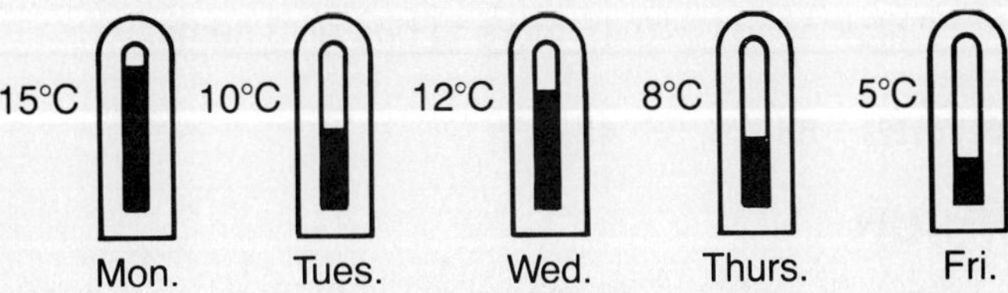

Solution

$$\begin{array}{r} 15 \\ 10 \\ 12 \\ 8 \\ +\ 5 \\ \hline 50 \end{array} \qquad \begin{array}{r} 10 \\ 5\overline{)50} \end{array}$$

The average morning temperature is 10°C.

133 Multiple-Step Problem

A baseball glove costs $25.00, a bat costs $10.95, and a ball costs $4.50. Derek has saved $30.00. How much more does he need to buy all 3 items?

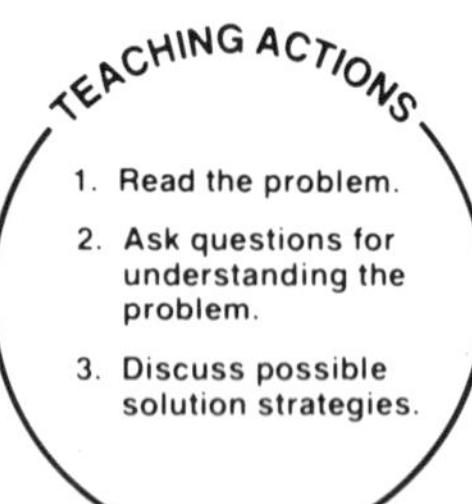

Understanding the Problem

- How much does the glove cost? ($25.00) The bat? ($10.95) The ball? ($4.50)
- How much has Derek saved? ($30.00)

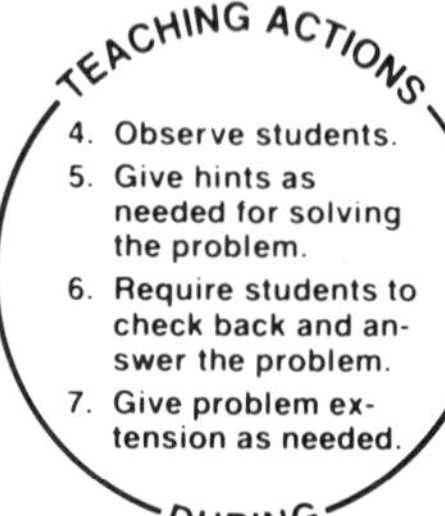

Planning a Solution

- What is the total cost of the glove, bat, and ball? ($25.00 + 10.95 + 4.50 = $40.45)
- Which operation would you use to find out how much more money Derek needs? (subtraction)

Finding the Answer

Choose the Operations

$$\begin{array}{r} \$25.00 \\ 10.95 \\ +\ 4.50 \\ \hline \$40.45 \end{array} \rightarrow \begin{array}{r} \$40.45 \\ -\ 30.00 \\ \hline \$10.45 \end{array}$$

Derek needs $10.45 more to buy all 3 items.

Related Problems: 88, 78, 58, 48, 43

Problem Extension

The store is having a sale. The glove costs $8.00 less than the original price, the bat costs $3.00 less than the original price, and the ball costs $0.50 less than the original price. Does Derek have enough money to buy the items now? (yes)

134 Process Problem

David's age this year is a multiple of 5. Next year, David's age will be a multiple of 7. How old is David now?

Understanding the Problem

- What do we know about David's age this year? (multiple of 5)
- What do we know about David's age next year? (multiple of 7)

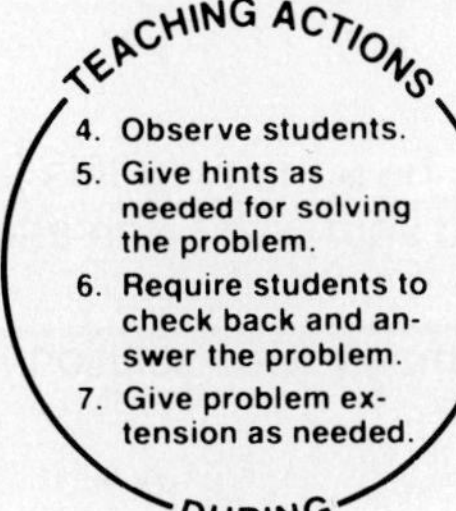

Planning a Solution

- List some multiples of 5. (5, 10, 15, . . .) Multiples of 7. (7, 14, 21, . . .)
- Guess what David's age might be this year and add 1 to it to see if that number is a multiple of 7. (See solution.)

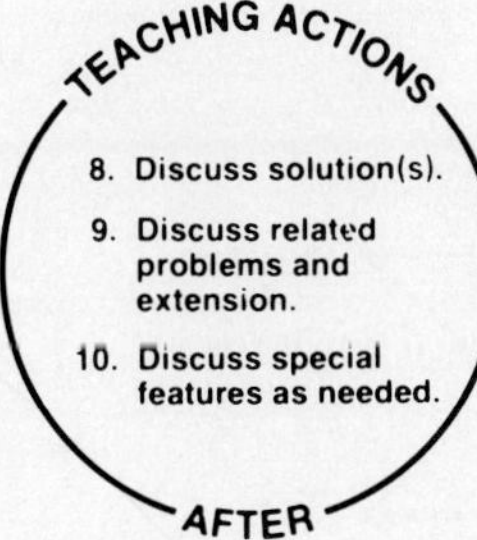

Finding the Answer

Guess and Check

- Try 10 and 11. (No, 11 is not a multiple of 7.)

and so on

Make an Organized List

Multiples of 5	5, 10, 15, (20)
Multiples of 7	7, 14, (21), 28

David is 20 years old now.

Related Problems: 129, 104, 100, 99, 95

Problem Extension

In how many more years will David's age be a multiple of both 5 and 7? (In 15 years, when he is 35.)

135 Process Problem

Cathy's mom bought her 3 new shirts that were red, blue, and green and 3 new pairs of jeans that were yellow, tan, and white. All the shirts could be worn with any of the jeans. How many different outfits could Cathy make?

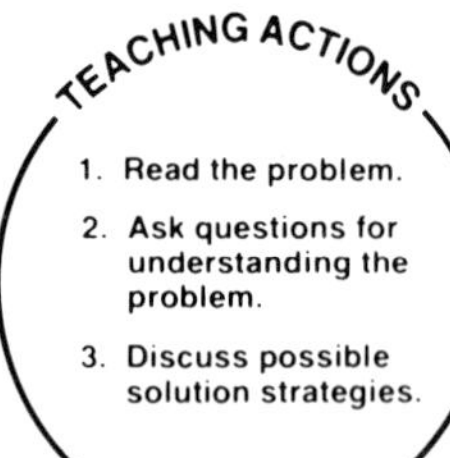

Understanding the Problem

- How many different shirts? (3) Jeans? (3)
- What is Cathy doing with the shirts and jeans? (She's making different outfits.)

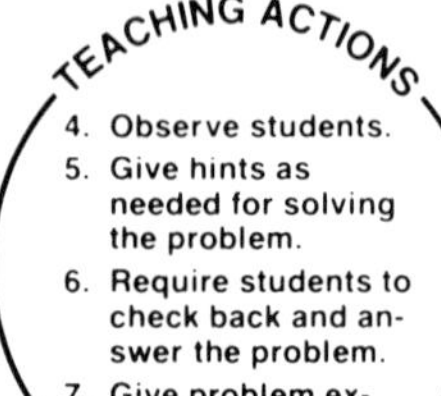

Planning a Solution

- If you match the red shirt with the yellow jeans, how many outfits is that? (1) What other jeans can you match the red shirt with? (tan and white)
- Make an organized list to show the outfits you can make. (See solution.)

Finding the Answer

Make an Organized List

Red—Yellow
Red—Tan
Red—White

Blue—Yellow
Blue—Tan
Blue—White

Green—Yellow
Green—Tan
Green—White

Draw a Picture

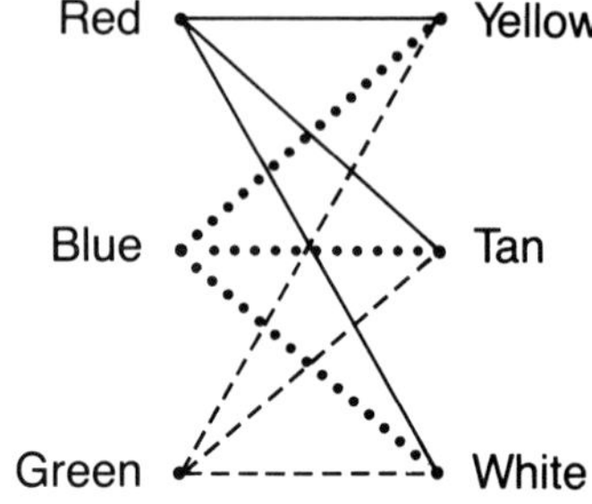

Cathy could make 9 different outfits.

Related Problems: 134, 110, 109, 100, 99

Problem Extension

How many different outfits could Cathy make if she added 3 different scarves? (27)

136 Skill Activity

These problems have the numbers omitted. Identify the operation or operations needed to solve each problem.

1. Karen has some money. She wants to buy a new pair of skis. How much more money will she need?

2. Ved arranged his baseball cards into several different teams. If he had the same number of cards for each team, how many baseball cards did he have?

Discussion

Possible answers:

1. Use subtraction to solve this problem.

2. Use multiplication to solve this problem.

137 One-Step Problem

Farmer Brown planted 225 tomato plants in 25 rows with the same number of plants in each row. If each plant produces 5 tomatoes, how many tomatoes will he get from his plants?

Solution

$$\begin{array}{r} 225 \\ \times\ \ 5 \\ \hline 1{,}125 \end{array}$$

He will get 1,125 tomatoes from his plants.

Note: The number of rows is unnecessary data.

138 Multiple-Step Problem

After the basketball game, Grace and Kevin had a late meal. Kevin ordered a chicken salad sandwich ($0.95), an order of cole slaw ($0.65), a piece of fruit pie ($0.47), and a glass of milk ($0.35). Grace ordered a chicken salad sandwich, an order of cole slaw, and a glass of milk. What was the total cost of Grace and Kevin's orders?

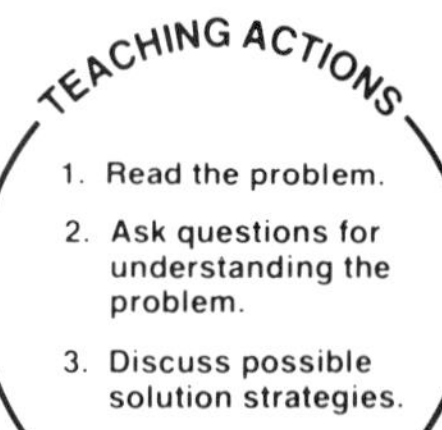

Understanding the Problem

- How much is a chicken salad sandwich? ($0.95) Cole slaw? ($0.65) Fruit pie? ($0.47) Glass of milk? ($0.35)
- What did Kevin order? (chicken salad sandwich, cole slaw, fruit pie, milk)
- What did Grace order? (chicken salad sandwich, cole slaw, milk)

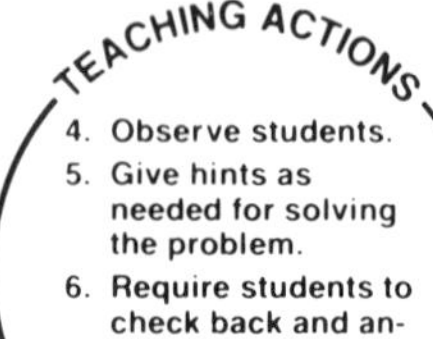

Planning a Solution

- What would you do to find the cost of Kevin's meal? (add)
- What numbers would you use to find the total of Grace's order? ($0.95 + $0.65 + $0.35)
- Which operation would you use to find the total cost of both orders? (addition)

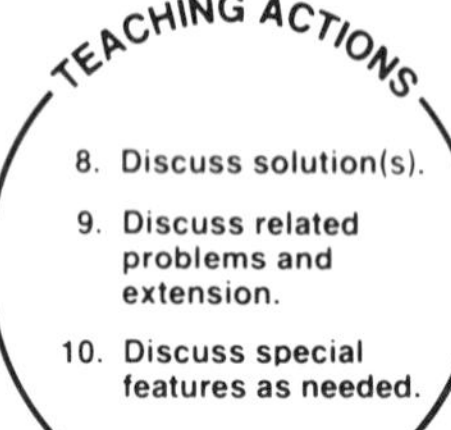

Finding the Answer

Choose the Operations

$$\begin{array}{r} \$0.95 \\ 0.65 \\ 0.47 \\ +\ 0.35 \\ \hline \$2.42 \end{array} \quad \rightarrow \quad \begin{array}{r} \$0.95 \\ 0.65 \\ +\ 0.35 \\ \hline \$1.95 \end{array} \quad \rightarrow \quad \begin{array}{r} \$2.42 \\ +\ 1.95 \\ \hline \$4.37 \end{array}$$

The total cost of Grace and Kevin's order was $4.37.

Related Problem: 83

Problem Extension

What was their change if they paid for the order with a $5.00 bill? ($0.63)

139 Process Problem

Mariko passed around a basket of strawberries to the girls at her party. There were 8 girls there. The first girl took 1 strawberry, the second girl took 3 strawberries, the third girl took 5 strawberries, and so on in that manner. After the last girl took her strawberries, the basket was empty. How many strawberries were in the basket in the beginning?

TEACHING ACTIONS

1. Read the problem.
2. Ask questions for understanding the problem.
3. Discuss possible solution strategies.

BEFORE

Understanding the Problem

- How many girls were at the party? (8)
- How many strawberries did the first girl take? (1) The second girl? (3) The third girl? (5)

TEACHING ACTIONS

4. Observe students.
5. Give hints as needed for solving the problem.
6. Require students to check back and answer the problem.
7. Give problem extension as needed.

DURING

Planning a Solution

- How many more strawberries did the second girl take than the first? (2)
- How many more strawberries did the third girl take than the second? (2)
- Make a table and look for a pattern. (See solution.)

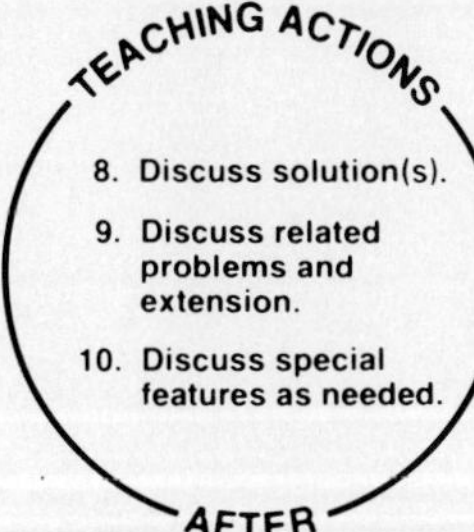

Finding the Answer

Make a Table/Look for a Pattern

Girl	1	2	3	4	5	6	7	8
Number of Strawberries Taken	1	3	5	7	9	11	13	15
Total Number Taken	1	4	9	16	25	36	49	64

Pattern: The number of strawberries taken increases by 2 with each girl.

There were 64 strawberries in the basket.

Related Problems: 129, 125, 124, 115, 114

Problem Extension

Mariko had eaten 5 strawberries and given 3 to a friend before the party. How many strawberries had been in the basket? (72)

140 Process Problem

Robert wanted an allowance. His father said he had a choice of getting it on a weekly or on a daily basis. He said he would either pay him $1.25 a week or pay him in the following manner each day for a week: On Monday he would give him $0.01; on Tuesday, $0.02; on Wednesday, $0.04; and so on through Sunday. Which way would Robert receive the higher allowance?

TEACHING ACTIONS

1. Read the problem.
2. Ask questions for understanding the problem.
3. Discuss possible solution strategies.

BEFORE

Understanding the Problem

- What would Robert's weekly allowance be? ($1.25)
- How was his father going to pay him on a daily basis? (Monday, $0.01; Tuesday, $0.02; Wednesday, $0.04; and so on)

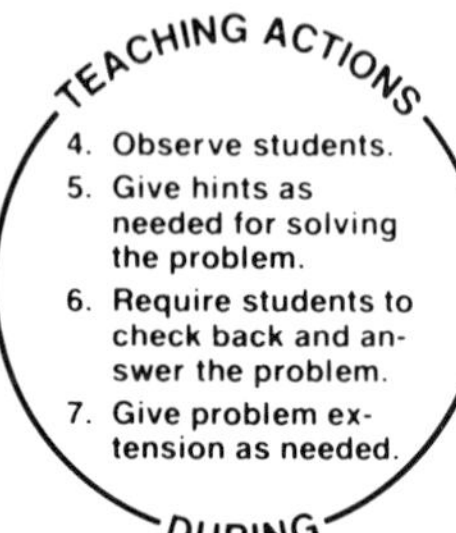

Planning a Solution

- How much will Robert get on Thursday? ($0.08) On Friday? ($0.16)
- Make a table to show how much Robert will get each day. (See solution.)

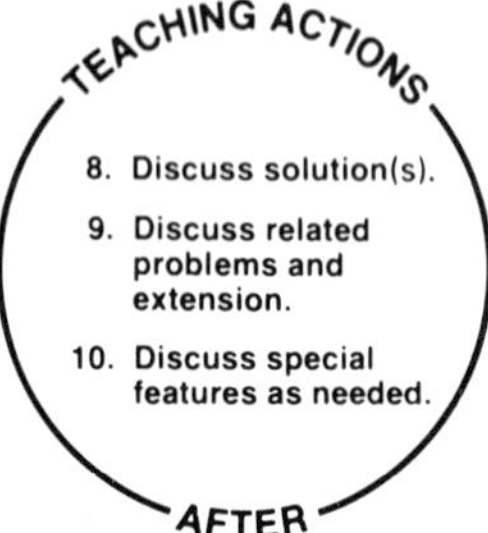

Finding the Answer

Make a Table

Day	*Monday*	*Tuesday*	*Wednesday*	*Thursday*	*Friday*	*Saturday*	*Sunday*
Amount	$0.01	$0.02	$0.04	$0.08	$0.16	$0.32	$0.64
Total	$0.01	$0.03	$0.07	$0.15	$0.31	$0.63	$1.27

Robert would get $0.02 more per week on a daily basis than he would on a weekly basis.

Related Problems: 139, 129, 125, 124, 115

Problem Extension

Suppose Robert's father offered either to give him $2.50 for 2 weeks or to continue this pattern of calculating a daily allowance for another week. How much would Robert have at the end of 2 weeks using the "doubling" approach? ($161.33)

141 Skill Activity

Write a story problem that this table would help you to solve:

International Phone Rates (First 3 min.)

	Station-to-Station	*Nights/Sundays*
Australia	**$7.80**	**$5.85**
Brazil	**7.80**	**5.85**
Finland	**6.30**	**5.10**
Hong Kong	**7.05**	**7.05**
Norway	**6.30**	**5.10**
United Kingdom	**4.80**	**3.75**

Discussion

Possible problem:

Last month I called my brother in Norway on a Sunday and my sister in Hong Kong on a weekday. Each call lasted 3 min. How much did I spend altogether for these two calls?

142 One-Step Problem

Kate jogs around the park each day. How far is it around the park?

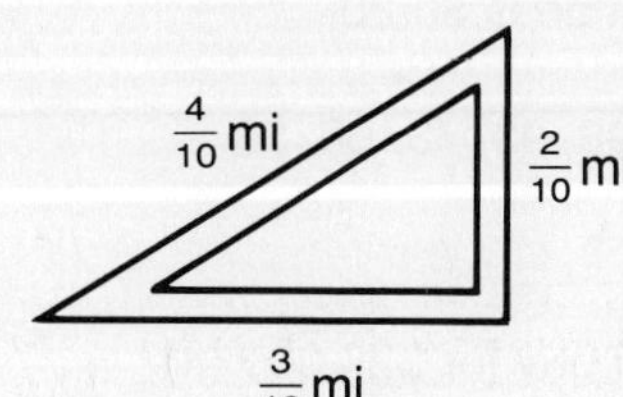

Solution

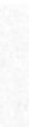

$$\begin{array}{r} \frac{4}{10} \\ \frac{3}{10} \\ + \frac{2}{10} \\ \hline \frac{9}{10} \end{array}$$

It is $\frac{9}{10}$ mi around the park.

143 Multiple-Step Problem

Peaches are on sale for \$0.40 a can. There are a dozen cans in a box. I want to buy a box of peaches and pay for it with a \$5.00 bill. What is my change?

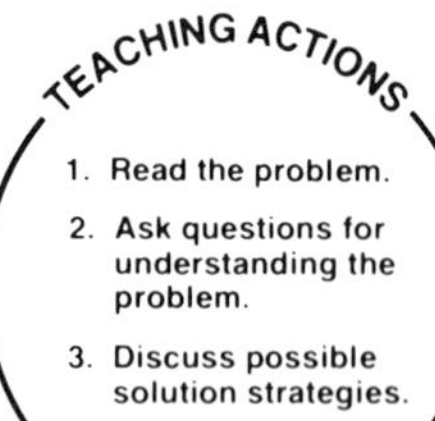

Understanding the Problem

- How much are peaches per can? (\$0.40)
- How many cans are there in a box? (a dozen, or 12)
- How much money do I have? (\$5.00)
- How many peaches do I want to buy? (1 box)

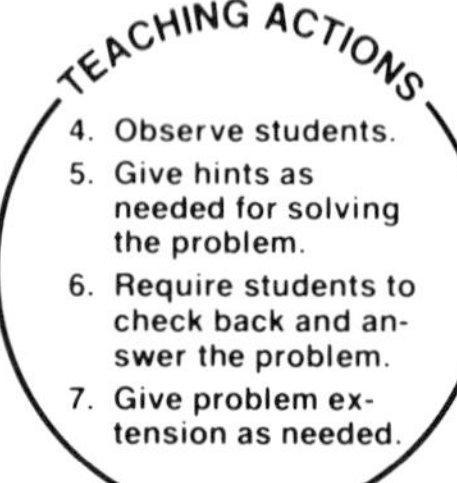

Planning a Solution

- If peaches cost \$0.40 a can and I want to buy a box, how much will that cost? ($\$0.40 \times 12 = \$4.80$)
- Which operation would you use to find the change? (subtraction)

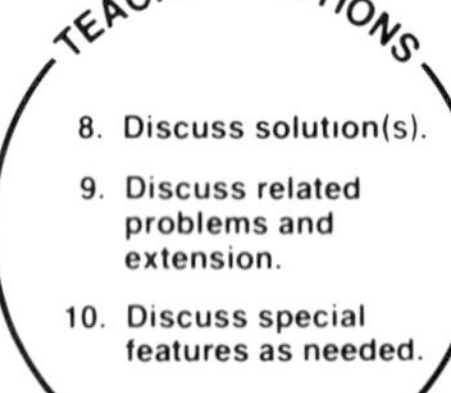

Finding the Answer

Choose the Operations

$$\begin{array}{r} \$0.40 \\ \times\ 12 \\ \hline \$4.80 \end{array} \quad \rightarrow \quad \begin{array}{r} \$5.00 \\ -\ 4.80 \\ \hline \$0.20 \end{array}$$

My change is \$0.20.

Related Problems: 98, 93, 73, 63, 38

Problem Extension

How many boxes of peaches can I buy for \$20.00? (4)

144 Process Problem

When Michael came home from Ellen's party, he told his mother about the 5 girls that were there. Barbara wore a red dress and Ellen wore a blue one. Mindy did not wear yellow. Sara and the girl in green won a contest against Barbara and the girl in yellow. He liked the girl in the brown skirt best. What color did Janice wear, and whom did Michael like best?

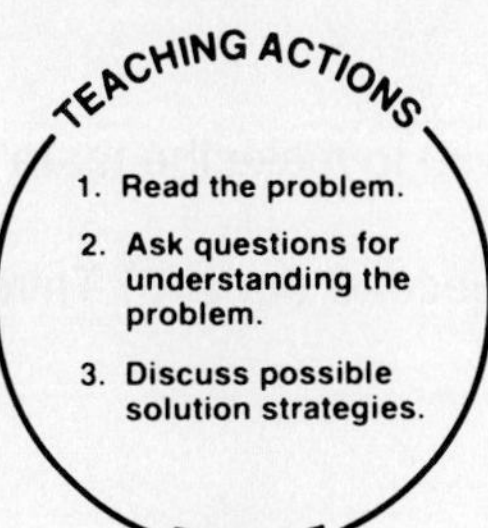

Understanding the Problem

- Who were the girls at the party? (Barbara, Ellen, Mindy, Sara, Janice)
- What colors were mentioned? (red, blue, yellow, green, brown)

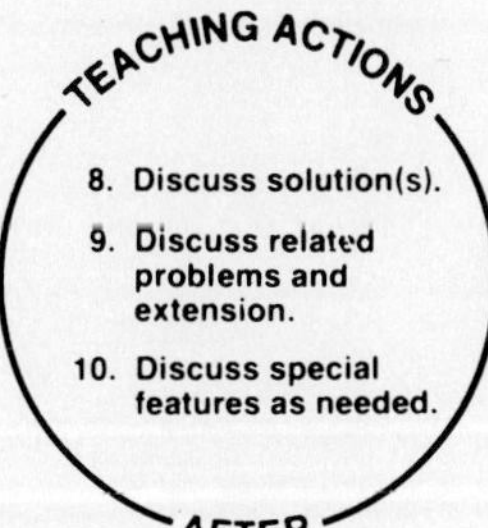

Planning a Solution

- What color did Barbara wear? (red) Ellen? (blue)
- What do we know about Mindy? (She did not wear yellow.) Sara? (She and the girl in green won a contest against Barbara and the girl in yellow.)
- Make a chart to help you use logical reasoning. (See solution.)

TEACHING ACTIONS
8. Discuss solution(s).
9. Discuss related problems and extension.
10. Discuss special features as needed.
AFTER

Finding the Answer

Make a Chart/Use Logical Reasoning

	Red	*Blue*	*Yellow*	*Green*	*Brown*
Barbara	X				
Ellen		X			
Mindy			No	X	
Sara				No	X
Janice			X		

Sara and the girl in green won a contest against Barbara and the girl in yellow.
Girl in green—Mindy
Girl in yellow—Janice

Janice wore yellow and Michael liked Sara best.

Related Problems: 109, 74, 70, 69, 35

Problem Extension

Suppose Michael made a mistake and Mindy was wearing yellow, Ellen was not wearing blue, and Janice was wearing green. What colors would each girl be wearing? (Barbara, red; Ellen, brown; Mindy, yellow; Sara, blue; Janice, green)

145 Process Problem

Chad was trying out for the swimming team. He had to be able to swim 30 laps by the end of the second week. He was not able to swim on weekends. On the first day he swam 1 lap; on the second day, 5 laps; on the third day, 9 laps; and so on. Was Chad able to swim enough laps at the end of 2 weeks to make the team?

TEACHING ACTIONS

1. Read the problem.
2. Ask questions for understanding the problem.
3. Discuss possible solution strategies.

BEFORE

Understanding the Problem

- How many laps did Chad have to be able to swim to make the team? (30)
- How many did Chad swim the first day? (1) Second day? (5) Third day? (9)

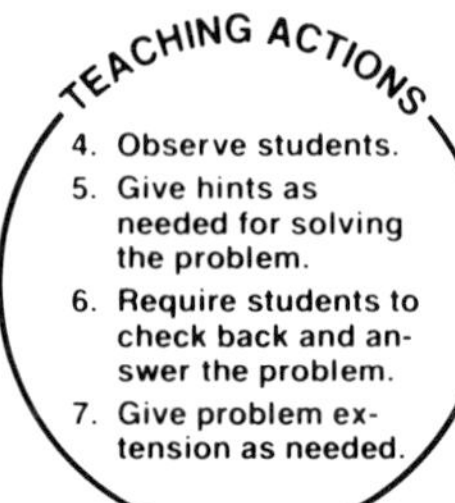

Planning a Solution

- How many more laps did he swim on the second day than on the first day? (4)
- How many more did he swim on the third day than on the second day? (4)
- Make a table and look for a pattern. (See solution.)

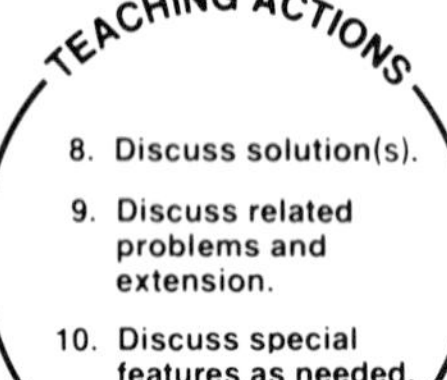

Finding the Answer

Make a Table/Look for a Pattern

Day	Number of Laps
Monday	1
Tuesday	5
Wednesday	9
Thursday	13
Friday	17
Monday	21
Tuesday	25
Wednesday	29
Thursday	33
Friday	37

← Greater than 30 (Thursday, 33)

Pattern: The number of laps increases by 4 each time.

Yes, Chad was able to swim enough laps to make the team.

Related Problems: 140, 139, 129, 125, 124

Problem Extension

Suppose Chad continues at this rate for another week (excluding weekends). How many laps will he then be swimming? (57)

146 Skill Activity

Determine whether the answers given for these problems are reasonable. If an answer is not reasonable, explain why.

1. Bonnie works 2 hours a day delivering the morning papers. If she works 6 days a week, how many hours does she work in 4 weeks?
   Answer: Bonnie works 500 hours in 4 weeks.

2. The 49¢ table tennis balls are packed in boxes holding 2 balls. The 30¢ table tennis balls are packed in boxes holding 3 balls. How many boxes are needed to hold 60 of the 30¢ table tennis balls?
   Answer: 20 boxes are needed.

Discussion

Possible answers:

1. This answer is not reasonable because $2 \times 6 = 12$ and $12 \times 4 = 48$. The correct answer is 48 hours.

2. This answer is reasonable because $60 \div 3 = 20$.

147 One-Step Problem

What is the distance from Keiko's house to Ken's house?

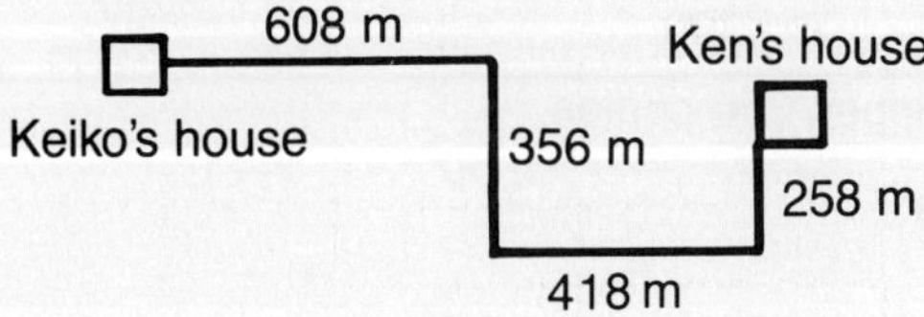

Solution

```
   608
   356
   418
 + 258
 -----
 1,640
```

The distance from Keiko's house to Ken's house is 1,640 m.

148 Multiple-Step Problem

The workers at the bottling plant were packing bottles for shipping. There were 8 bottles in a carton and 3 cartons in a case. If they shipped 15 cases of bottles, how many bottles did they ship?

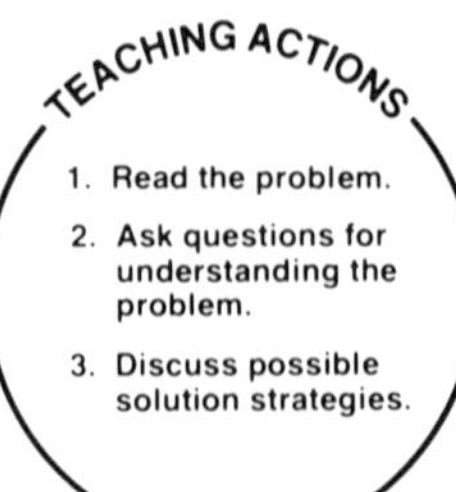

Understanding the Problem

- How many bottles are in a carton? (8)
- How many cartons are in a case? (3)
- How many cases were shipped? (15)

Planning a Solution

- If there are 8 bottles in a carton and 3 cartons in a case, how many bottles are there in a case? (24)
- If you know the number of cases that were shipped and the number of bottles in each case, which operation would you use to find out how many bottles were shipped? (multiplication)

TEACHING ACTIONS

8. Discuss solution(s).
9. Discuss related problems and extension.
10. Discuss special features as needed.

AFTER

Finding the Answer

Choose the Operations

$$\begin{array}{r} 8 \\ \times\ 3 \\ \hline 24 \end{array} \quad \rightarrow \quad \begin{array}{r} 24 \\ \times\ 15 \\ \hline 120 \\ 24 \\ \hline 360 \end{array}$$

They shipped 360 bottles.

Related Problem: 108

Problem Extension

Suppose 2 cases fell off along the way and 13 bottles were broken when the truck hit a hole in the road. How many bottles arrived safely? (299)

149 Process Problem

Adam, Ruby, Teresa, Pete, Angie, and Sam were chosen for infield positions on the baseball team. Sam plays shortstop. The pitcher and catcher are both girls. Angie is the catcher. Pete does not play second or third base. Adam plays third base. Ruby and the pitcher are sisters. Find the position played by each team member.

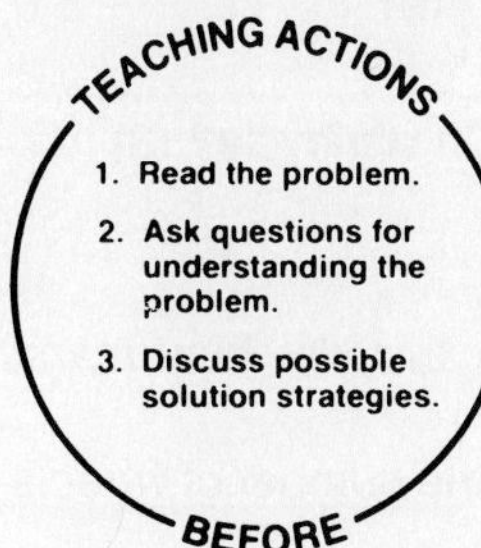

Understanding the Problem

- Who were chosen for positions in the infield? (Adam, Ruby, Teresa, Pete, Angie, Sam)
- What positions were they chosen for? (pitcher, catcher, first base, second base, third base, shortstop)
- Do we know the positions of any team members? (Yes, Sam plays shortstop, Angie is the catcher, Adam plays third base.)

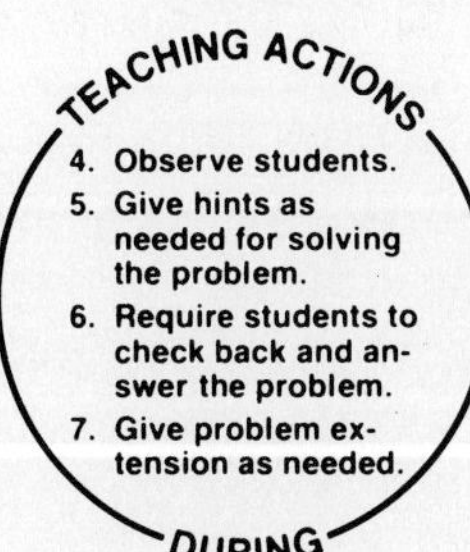

Planning a Solution

- What do we know about the pitcher and catcher? (both girls)
- Since Angie is the catcher and Ruby and the pitcher are sisters, who is the pitcher? (Teresa)
- What do we know about Pete? (He does not play second or third base.)
- Make a chart to help you use logical reasoning. (See solution.)

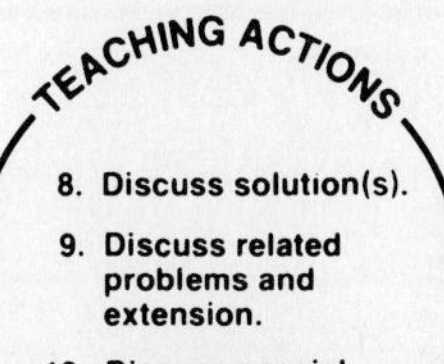

Finding the Answer

Make a Chart/Use Logical Reasoning

	Catcher (girl)	Pitcher (girl)	First	Second	Third	Shortstop
Adam					X	
Ruby		Not		X		
Teresa		(Ruby's sister) X				
Pete			X	Not	Not	
Angie	X					
Sam						X

Angie is the catcher, Teresa is the pitcher, Ruby plays second, Pete plays first, Adam plays third, and Sam is the shortstop.

Related Problems: 144, 109, 74, 70, 69

Problem Extension

Suppose the pitcher does not have to be a girl and Ruby and the person playing second are sisters. What would the positions be? (Adam—third, Ruby—first, Teresa—second, Pete—pitcher, Angie—catcher, Sam—shortstop)

150 Process Problem

In Booth's Bicycle Shop, Henry counted some bicycles and tricycles. He counted 19. When he counted the wheels, he got a total of 45. How many bicycles and tricycles were there?

TEACHING ACTIONS

1. Read the problem.
2. Ask questions for understanding the problem.
3. Discuss possible solution strategies.

BEFORE

TEACHING ACTIONS

4. Observe students.
5. Give hints as needed for solving the problem.
6. Require students to check back and answer the problem.
7. Give problem extension as needed.

DURING

TEACHING ACTIONS

8. Discuss solution(s).
9. Discuss related problems and extension.
10. Discuss special features as needed.

AFTER

Understanding the Problem

- How many bicycles and tricycles did he count? (19)
- How many wheels did he count? (45)
- How many wheels are there on a bicycle? (2) On a tricycle? (3)

Planning a Solution

- If you counted the wheels on 1 bicycle and 1 tricycle, how many wheels would you count? (2 + 3 = 5)
- Make a table showing bicycles and tricycles and the number of wheels on them. (See solution.)
- Try making a guess. Then check your guess. (See solution.)

Finding the Answer

Make a Table/Look for a Pattern

Bicycles	Wheels
1	2
2	4
3	6
4	8
5	10
6	12
7	14
8	16
9	18
10	20
11	22
(12)	(24)
13	26
14	28
15	30
16	32
17	34
18	36
19	38

Tricycles	Wheels
1	3
2	6
3	9
4	12
5	15
6	18
(7)	(21)
8	24
9	27
10	30
11	33
12	36
13	39
14	42
15	45

Guess and Check

- Try 10 bicycles, 9 tricycles—20 + 27 = 47. (no)
- Try 11 bicycles, 8 tricycles—22 + 24 = 46. (no)
- Try 12 bicycles, 7 tricycles—24 + 21 = 45. (yes)

There were 12 bicycles and 7 tricycles in Booth's Bicycle Shop.

Related Problems: 145, 140, 139, 134, 129

Problem Extension

In another corner were two-wheeler dirt bikes and smaller bikes with training wheels. Altogether there were 19 bikes. If Henry counted 56 wheels, how many of each kind were there? (10 two-wheelers and 9 four-wheelers)